CLIMATE CHANGE AND FOOD SYSTEMS RESILIENCE IN SUB-SAHARAN AFRICA

Edited by
Lim Li Ching, Sue Edwards and Nadia El-Hage Scialabba

Food and Agriculture Organization of the United Nations | 2011

The designations employed and the presentation of material in this information product do not imply the expression of any opinion whatsoever on the part of the Food and Agriculture Organization of the United Nations (FAO) concerning the legal or development status of any country, territory, city or area or of its authorities, or concerning the delimitation of its frontiers or boundaries. The mention of specific companies or products of manufacturers, whether or not these have been patented, does not imply that these have been endorsed or recommended by FAO in preference to others of a similar nature that are not mentioned.

ISBN 978-92-5-106876-2

Cover photo and African artcrafts presented in this book have been kindly provided by the personal archive of Marzio Marzot.

PREFACE

This volume is published at a critical juncture in the history of human impact on the biosphere – a history we can trace back some 10 000 years – to the era of the agricultural revolution. Now, in retrospect, we see a revolution that was indeed slow and by today's definition, hardly a revolution. Nevertheless, it was fast enough to feed a growing population and see it through several food crises. But it also was a time during which the rate of species extinction rose dramatically and changed the natural environment. This loss of the genetic wealth of our biosphere was slightly compensated by the agricultural biodiversity that humans ingeniously developed in their fields which ultimately spread and evolved into components of our diverse agricultural ecosystems.

Only lately, just two or three centuries ago, came the more fast-paced industrial revolution and its machines that took on many of humanity's tasks – machines powered by an ever-increasing burning of fossil fuels. As a consequence of this fuel-driven power, the greenhouse gas content of the atmosphere increased as did the temperature of the biosphere.

The fast rate of changing environmental conditions, namely temperature and rainwater levels and variability, could no longer be matched by species adaptation.

The industrial revolution also led to mass production, because of the ease with which the new machines could spew out identical copies of tools and furniture and cars, more efficiently and cheaply than handmade versions. In following decades, this acceptance of homogeneity spread from factories into fields. Agriculture's focus shifted to commercial crop varieties with their improved output, at the loss of those "handmade" or indigenous varieties adapted to unique environments and resistant to pests and diseases. The result: erosion of agricultural biodiversity which, in turn, has weakened the ability of agro-ecosystems to adapt to climate change and increased weather variability.

There is absolutely no doubt that there is need to increase global food production to meet the demand of a growing population. However, the critical issue in increasing production is recognizing the need to support, rather than disrupt, the nutrient and energy dynamics and the biodiversity of agricultural systems and of ecosystem services as a whole. These dynamics and this diversity must be used to maximize the efficiency of energy use and nutrient cycling in both agricultural and non-agricultural systems. Today, due to climate change and volatile fossil fuel prices, the over-simplified solution of replacing lost nutrients with agrochemicals that are themselves derived from fossil fuels is no longer viable.

The need to create resilient and productive agro-ecosystems can be met by starting with biomass-rich soils, which have the potential to sequester carbon. But that is just part of the story. The biomass also improves soil structure which means better moisture-holding capacity, nutrient retention and, ultimately, reduced vulnerability to water and wind erosion. In addition, the soil's improved capacity to retain nutrients and water has a direct positive effect on crops and other produce, and the coverage offered by rigorous vegetation reduces wind erosion. In the big picture, healthy agro-ecosystems mean improved environmental well being as well as increased yields.

This volume, *Climate Change and Food Systems Resilience in Sub-Saharan Africa*, demonstrates the possibility of harmonizing agricultural production with the wellbeing of the biosphere – and that this can be achieved in Africa, our biosphere's least developed continent, and the continent which is likely to suffer most from climate change.

The work presented in this volume stems from a Conference on Ecological Agriculture held in Ethiopia in 2008. Through the discussions held during this Conference and field visits to Tigray, a region struck by hunger in the eighties and largely food secure today, participants shared insights on Africa's potential for intensifying its agriculture through a better use of natural resources and ecosystem services. This volume represents the collective knowledge and subsequent writings of this Conference' participants.

The different chapters capitalize on assessments and experiences such as: lessons learned from Asia's Green Revolution on agricultural communities; trends in African agricultural knowledge, science and technology; trade policy impacts on food production; conditions for success of water interventions for the African rural poor; and climate change implications for agriculture and food systems. Case studies share the practical experiences, lessons and successes from across Africa, demonstrating that it is possible to produce food sufficiently and at the same time, care for the biosphere. The chapters documenting Tigray's experience in rehabilitating watersheds for local food security show us how social and environmental goods and services go hand in hand.

The chapters on organic agriculture and fair trade in West Africa and Uganda demonstrate that the impacts on food security and trade need not to be negative. The chapters depicting smallholder practices for making compost and maintaining local seed supplies under low-input conditions offer guidance to communities seeking resilience.

In short, this book offers hope – hope that we can continue to produce – enough nutritious food to enable us, our children and our grandchildren to live healthy lives and that our biosphere can exist in harmony with the life that it has generated.

Tewolde Berhan Gebre Egziabher
Director-General
Environmental Protection Authority
Addis Ababa, Ethiopia

Alexander Müller
Assistant Director-General
Natural Resources Management and Environment Department
Food and Agriculture Organization of the United Nations
Rome, Italy

EDITORIAL NOTE

This book was edited by Lim Li Ching, Third Work Network (Malaysia), Sue Edwards, Institute for Sustainable Development (Ethiopia) and Nadia El-Hage Scialabba, FAO (Italy). The editors wish to thank all authors who have contributed to this volume, as well as the Swedish International Development Cooperation Agency (SIDA) for its generous financial support.

CONTENTS

CLIMATE CHANGE AND FOOD SYSTEMS RESILIENCE IN SUB-SAHARAN AFRICA

A MANUAL FOR THE PREPARATION OF WOREDA AND LOCAL
COMMUNITY PLANS FOR ENVIRONMENTAL MANAGEMENT FOR
SUSTAINABLE DEVELOPMENT
page 199

SUCCESSES AND CHALLENGES IN ECOLOGICAL AGRICULTURE:
EXPERIENCES FROM TIGRAY, ETHIOPIA
page 231

ADOPTION OF ORGANIC FARMING TECHNOLOGIES:
EVIDENCE FROM A SEMI-ARID REGION IN ETHIOPIA
page 297

ORGANIC AGRICULTURE AND FAIR TRADE IN WEST AFRICA
page 323

ORGANIC TRADE PROMOTION IN UGANDA:
A CASE STUDY OF THE EPOPA PROJECT
page 347

ESTABLISHING A COMMUNITY SUPPLY SYSTEM:
COMMUNITY SEED BANK COMPLEXES IN ETHIOPIA
page 361

HOW TO MAKE AND USE COMPOST
page 379

CONFERENCE ON ECOLOGICAL AGRICULTURE: MITIGATING CLIMATE CHANGE, PROVIDING FOOD SECURITY AND SELF-RELIANCE FOR RURAL LIVELIHOODS IN AFRICA

African Union Headquarters
Addis Ababa, Ethiopia
26-28 November 2008

CONTENTS

ANNEX

INTRODUCTION

The Conference on Ecological Agriculture: Mitigating Climate Change, Providing Food Security and Self-Reliance for Rural Livelihoods in Africa was held in Addis Ababa, Ethiopia on 26-28 November 2008. It was organised by the African Union (AU), Food and Agriculture Organization of the United Nations (FAO) and the Ministry of Agriculture and Rural Development of Ethiopia, in collaboration with the Institute for Sustainable Development (ISD), Ethiopia and the Third World Network (TWN).

Over 80 participants from 15 African countries - Benin, Burundi, Djibouti, Ethiopia, Kenya, Madagascar, Malawi, Mali, Nigeria, Rwanda, Sudan, Tanzania, Uganda, Zambia and Zimbabwe - attended the Conference. The participants included policy-makers, agriculture experts representing governments, NGOs, farmers' organizations, universities, and international and regional bodies such as the AU, FAO, United Nations Conference on Trade and Development (UNCTAD), the UNEP-UNCTAD Capacity Building Task Force on Trade, Environment and Development (CBTF), International Food Policy Research Institute (IFPRI), International Assessment on Agricultural Knowledge, Science and Technology for Development (IAASTD) and World Food Programme (WFP).

The Conference was preceded by a field visit to the Axum area in Tigray Region in northern Ethiopia on 23-25 November 2008, to visit some of the communities of smallholder farmers that the Tigray Regional Bureau of Agriculture and Rural Development of Ethiopia and ISD have been working with on ecological agriculture since 1996. This was an appropriate experience to help focus attention on the aspects of the ecosystem that can easily respond to appropriate management, so as to stimulate discussion on experiences relevant for raising agricultural production, mitigating and adapting to climate change, and achieving the Millennium Development Goals (MDGs) in Africa.

The following are among the significant views, conclusions and recommendations expressed by participants during the Conference.

GENERAL VIEWS

The Conference heard several presentations and discussed the challenges facing African agriculture, not least among them the global food crisis, climate change and the conflicts with inappropriate biofuels development. Moreover, land degradation and the consequential loss of soil fertility, which are exacerbated by pests and erratic rainfall associated with climate change, are major constraints to improving agricultural production in Africa. Consequently, many local communities in African countries are food insecure. Trade policies also have implications for African food security and rural development, which need to be addressed, to stop the worrying trend of food import dependency and increasing vulnerability to external shocks.

The steep rise in petroleum prices and the consequent increase in the cost of chemical fertilizers and pesticides are making it essential to improve soil fertility and agricultural productivity in Africa through effective management of the local resources that are found in the agricultural and surrounding ecosystems. Many diverse and creative ecological agriculture (including organic agriculture) practices based on rich traditional knowledge and agrobiodiversity are found in Africa. Where supported by appropriate research and policy, it has been shown that these have been effective in tackling poverty and improving livelihoods.

In addition, this opens up the opportunity for Africa's smallholder farmers to become recognized as organic farmers producing for the growing global market fetching fair prices for their products. The global organic market growth has been about 15 percent per year over the past decade. Internal markets for organic products are also developing rapidly, particularly where consumers are made aware of the improvements to health from eating organic food.

The Conference heard presentations on the potential of ecological agriculture, including organic agriculture, to meet food security needs in Africa. Concrete examples and lessons learnt were presented from several African countries on practices that have successfully increased productivity and yields of crops, provided ecologically sound pest, weed and disease control, resulted in better water availability, met household and local food security needs, increased household income and improved livelihood opportunities, especially for women who are the majority of Africa's farmers. Other presentations focused on the potential of ecological agriculture to mitigate climate change, and to provide farmers with the means to adapt to climate change.

Participants discussed the need for appropriate national policies to support and build the capacity of farmers and agricultural professionals to implement and mainstream ecological/organic agriculture in Africa. Some of the major barriers and challenges to a transition to ecological agriculture were identified, and recommendations for charting the way forward in terms of policies, action plans and regional and international cooperation were made.

MAIN CONCLUSIONS

1. Ecological agriculture holds significant promise for increasing the productivity of Africa's smallholder farmers, with consequent positive impacts on food security and food self-reliance. This is demonstrated by efforts such as the Tigray Project, now working with over 20 000 farming families in Ethiopia, where crop yields of major cereals and pulses have almost doubled using ecological agricultural practices such as composting, water and soil conservation activities, agroforestry and crop diversification. Although Tigray was previously known as one of the most degraded Regions of Ethiopia, yet over the 12 years of the introduction and expansion of ecological agriculture, the use of chemical fertilizers has steadily decreased while total grain production has steadily increased.

2. As most poor farmers, particularly in degraded lands and in market-marginalised areas, are not able to afford external inputs, the principles and approach of the Tigray Project, based on ecological agriculture, offer farmers and their families a real and affordable means to break out of poverty and achieve food security, provided that relevant government commitment, support and capacity-building is provided to them.

3. Ecological agriculture also provides many other benefits, including to the environment, such as addressing land degradation and reducing the use of polluting chemical inputs, with consequent beneficial health impacts. Ecological agriculture helps foster agrobiodiversity and other essential environmental services, which improve agroecosystem resilience, helping farmers to better face risks and uncertainties. The productivity and diversity of crops also increase incomes and improve rural livelihoods.

4. Ecological agriculture has high climate change mitigation potential; for example avoiding the use of synthetic fertilizers results in reduced greenhouse gas emissions, particularly nitrous oxide. Ecological agriculture practices such as using leguminous crops, crop residues, cover crops and agroforestry enhance soil fertility and lead to the stabilization of soil organic matter and in many cases to a heightened sequestration of carbon in the soils.

5. Ecological agriculture assists farmers in adapting to climate change by establishing conditions that increase agroecosystem resilience to stress. Increasing an agroecosystem's adaptive capacity allows it to better withstand climate variability, including erratic rainfall and temperature variations and other unexpected events. Drawing on strong local community and farmers' knowledge and agrobiodiversity, ecological agriculture improves soil quality by enhancing soil structure and its organic matter content, which in turn promotes efficient water use and retains soil moisture. Such conditions simultaneously enhance soil conservation and soil fertility, leading to increased crop yields.

6. The development and growing of biofuels should not compete with food and other crops, and thus require comprehensive impact assessments. Locally-controlled bioenergy production that makes use of agricultural waste and biomass, such as through biogas digesters, could provide sustainable energy generation.

7. Food and energy demand and climate change are inducing land use changes and land access issues, which threaten the viability of farming and rural livelihoods. The resilience of agroecosystems can only be built by empowering

local communities, particularly women, to rehabilitate, adapt and improve their natural resource base for continued productivity, and by giving them the appropriate legal backing.

8. The implementation and scaling up of ecological agriculture face several constraints, including the lack of policy support at local, national, regional and international levels, resource and capacity constraints, and a lack of awareness and inadequate information, training and research on ecological agriculture at all levels.

RECOMMENDATIONS
Policy and planning

1. The AU and other regional organizations (e.g. Southern African Development Community, SADC; Economic Community of West African States, ECOWAS; Common Market for Eastern and Southern Africa, COMESA) are urged to take action to assist African governments in implementing policies and action plans on ecological agriculture. The AU Commission should also develop strategic partnerships with civil society and other actors to promote and implement ecological agriculture in the continent.

2. The FAO is called to assist the AU in developing an African Action Plan on Ecological Agriculture that will guide member countries in implementing relevant policies and action plans, as a matter of urgency.

3. Governments are urged to conduct in-depth assessments of agricultural conditions and policies in their countries, identify barriers to a transition to ecological agriculture and gaps in policy, and to ensure policy coherence such that ecological agriculture is promoted and facilitated. Meaningful impact of development actions also requires the extensive deployment of extension officers and direct involvement of local communities. Resources from the national, regional and international levels, including climate-related funds, should be made available to assist governments to implement policies and action plans on ecological agriculture.

4. Trade policies should be crafted so that they are supportive of ecological/organic agriculture. Governments are urged to ensure that commitments made at the multi-lateral and bilateral levels provide enough policy space to enable support for the agriculture sector, expansion of local food production, and effective instruments to provide local and household food security, farmers' livelihoods and meet rural development needs.

5. Governments are urged to provide support in linking farmers to markets, in the development of domestic and regional markets for organic agricultural products, and in assisting farmers to access regional and international markets. Building awareness on the environmental and health benefits of organic products, and creating linkages between producers and consumers through short supply chains for ecological produce, are needed in order to stimulate local demand and local markets.

Research and development

6. Institutions involved in ecological agriculture are requested to pool their expertise and identify ways to establish an African Centre of Excellence on Ecological Agricultural Research. Research priorities along the value chain, including key food crops and animals, best practices, economic aspects, main problems and solutions to these problems, should be identified in a participatory manner. Farmers' knowledge is a basic and important component of the research/ development continuum and research from the scientific community can complement and build on this.

7. The Conference participants agreed to establish a resource centre on ecological agriculture (e.g. an electronic library) to document best practices (including local knowledge and skills) and enable better communications, sharing of information and experiences on ecological agriculture.

8. Governments are urged to develop awareness, training and educational materials and curricula on ecological agriculture, including for students in schools, tertiary

educational institutions, graduate schools, extension officers and farmers. There is a need to include the mass media in awareness-raising efforts and to encourage consumers to appreciate the values of local organic products. Guidelines must be developed for training of trainers (e.g. extension officers) on watershed environmental management strategies and climate change adaptation practices through ecological agriculture. Improving soil fertility in dry and poorly vegetated areas must be given specific consideration.

Demonstration projects and technical assistance

9. Pilot projects on ecological agriculture should be established in each country to demonstrate the benefits of ecological agriculture to food security and rural livelihoods. In locations where Green Revolution projects are being launched or implemented, ecological agriculture pilot projects should be given the same financial and other kinds of support in order to allow comparative assessment of the two management systems' performance, including periodic documentation, monitoring and evaluation of impacts over the short-, medium- and long-terms. Where ecological agriculture projects already exist, they should be scaled up so as to encourage a wider impact on the environment and uptake by rural communities.

10. The training and technical assistance needs in relation to ecological agriculture should be identified and a list of experts compiled and targeted for continued capacity-building and training.

Implementation

11. The international community and African regional and national organizations, including the co-organisers of the Conference, are urged and encouraged to undertake follow-up activities, including providing policy and technical assistance to African governments, particularly the Ministries of Agriculture, Environment and Trade, in order to vigorously support ecological/organic

agriculture plans and programmes. Efforts should also include assisting Governments to tap climate-related funds in order to support capacity-building work in ecological agriculture.

12. The donor community is called upon to provide the resources required for ecological agriculture interventions to meaningfully support food security and rural livelihoods. This entails ensuring that adequate and balanced financial allocations are made for ecological agriculture projects. This is especially needed as heavy investments in industrial and chemically-oriented agriculture create disincentives to other agricultural management alternatives.

13. The Conference participants established a Standing Committee on Ecological Agriculture (see Annex A), which includes representatives from each participating country, in order to continue sharing experiences, enhance networking, undertake follow-up activities such as national workshops, and further the implementation of ecological/organic agriculture in their respective countries and at regional and international levels.

ANNEX A
STANDING COMMITTEE ON ECOLOGICAL AGRICULTURE

AFRICAN UNION (AU)
Dr Sarah Olembo
Senior Advisor
Department of Rural Economy
and Agriculture
Addis Ababa

INTERNATIONAL FEDERATION OF ORGANIC AGRICULTURE MOVEMENTS (IFOAM)
Hervé Bouagnimbeck
Africa Office Coordinator
IFOAM Head Office
Bonn, Germany

BENIN
Valery Lawson
Secretary General
NGO JINUKUN / COPAGEN
Cotonou
Mikpon Toussaint
Researcher
National Agricultural Research
Institute of Benin
Cotonou

BURUNDI
Annick Seziber
Legal Representative
Collectif des association Paysannes
pour l'auto development (CAPAD)
Bujumbura
Pascal Baridomo
Director Inades-Formation Burundi
Bujumbura

DJIBOUTI
Houssein Rayleh
Director of Djibouti Nature/
Coordinator of the Horn of Africa
Regional Environment Network
Djibouti
Chamake Mohamed Youssouf
Djibouti Government employee

ETHIOPIA
Sue Edwards
Director
Institute for Sustainable Development
Addis Ababa
Gebremedhin Birega
Manager
Eco Consumers' Association
of Ethiopia
Addis Ababa

KENYA
Monica Mueni
Assistant Director of Agriculture
Nairobi
Zacharia Makanya
Participatory Ecologic land Use
Management (PELUM)-Kenya
Country Coordinator
Thika

MALAWI
Esther Kamlongera
Counsellor
Malawi Embassy
Addis Ababa

MALI
Niaba Teme
Plant Breeder
Researcher
Mali Institute of Rural Economy
Bamako
Salikou Sanogo
Coordinator of Institute for
Research & the Promotion of
Alternatives in Development
Ministry of Agriculture
Planning and Statistics Department
Bamako

NIGERIA
Olugbenga AdeOluwa
Department of Agronomy,
Faculty of Agriculture and Forestry
University of Ibadan/ National Secretary
Network of Organic Agriculture
in Nigeria (NOAN)
IFOAM Contact Point
Coordinator for Nigeria
Ibadan

RWANDA
Aloys Semakuza
Coordinator of Bureau d'Appui aux
Initiatives Rurales (BAIR)
Gisenyi

SOMALILAND
Khadra Omer
Chairperson
Bawaaqo Voluntary Organization
(BVO)
Hargeisa

SUDAN
Sumaia Elsayed
Ahfad University for Women
Umdurnaman
Hayat Ahmed Elmahi
Sudanese Women General Union
(SWGU)
Khartoum

TANZANIA
Bashiru Abdul Hasani
Senior Program Officer
AGENDA
(Action for Environment & Development)
Dar es Salaam
Ombaeli Lemweli
Principal Economist
National Food Security Division
Ministry of Agriculture, Food Security
and Cooperatives
Dar es Salaam

UGANDA
Charles Ssekyewa
Director of Research
Ugandan Martyrs University
Kampala
Doleera Jackson
Executive Director
Development of the
Rural Economy (DERC)
Masindi

ZAMBIA
Kusiyo Mbikusita Lewanika
Executive Director
Lyambai Institute of Development
Mongu
Bernadette Lubozhya
Smallholder farmer and
Board member of Organic Producers &
Processors Association of Zambia
(OPPAZ)
Lusaka

ZIMBABWE
Medicine Masiiwa
Director, AIPAD Trust, Harare
Mukura Tamuka
Economist, Ministry of Agriculture
Harare

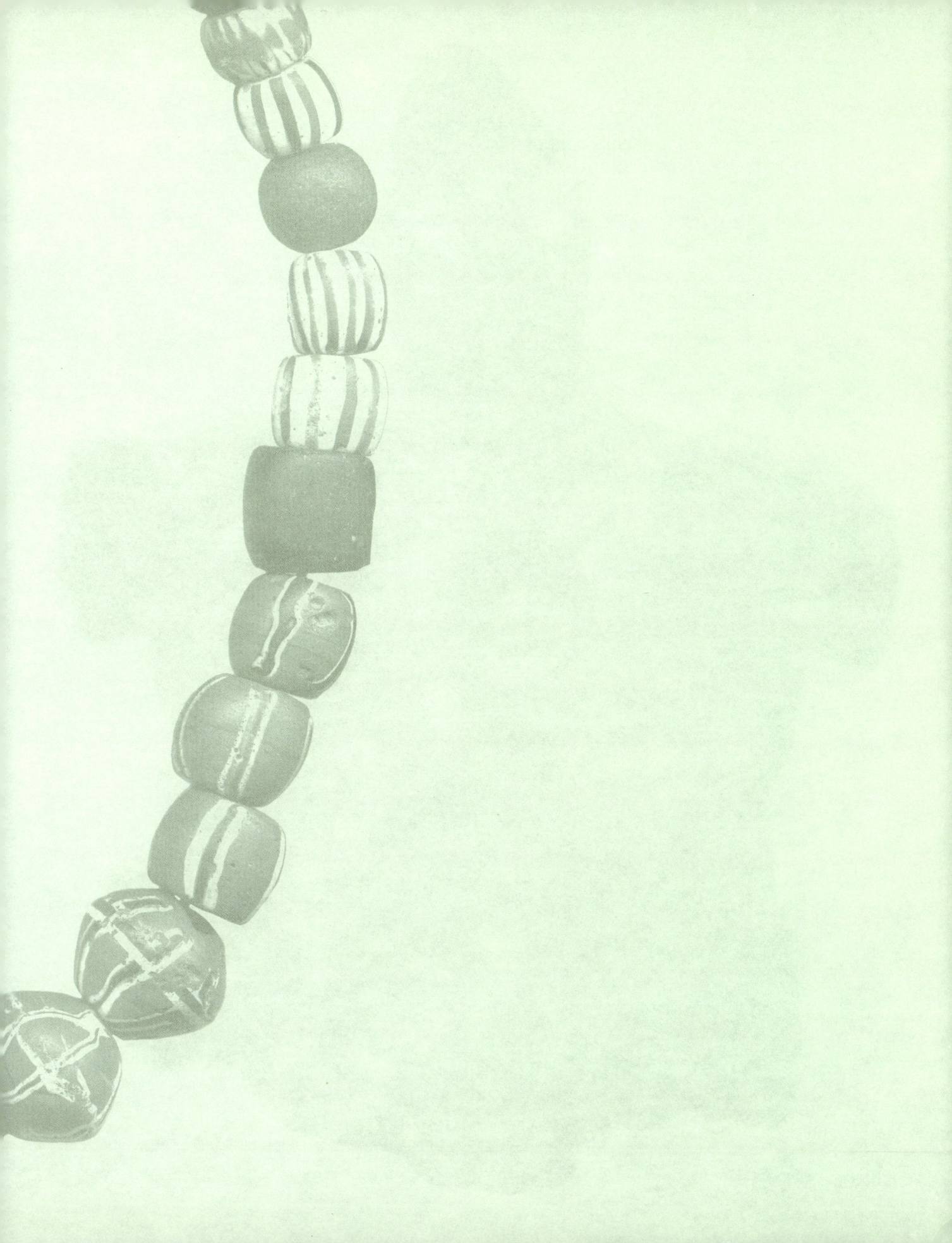

AFRICA'S POTENTIAL FOR THE ECOLOGICAL INTENSIFICATION OF AGRICULTURE

Tewolde Berhan Gebre Egziabher and
Sue Edwards

CONTENTS

FIGURES

TABLES

ABOUT THE AUTHORS

Tewolde Berhan Gebre Egziabher is the Director General of the Environmental Protection Authority of the Federal Democratic Republic of Ethiopia. He is a plant ecologist who has been a university academic staff member and an academic administrator as well as a negotiator on biodiversity issues.

Sue Edwards is the Director of the Institute for Sustainable Development (ISD), Addis Ababa, Ethiopia, and has been co-editor of the eight-volume "Flora of Ethiopia and Eritrea" since 1984. She is a taxonomic botanist, teacher and science editor by profession.

INTRODUCTION

The present ability or otherwise of Africa to cope with climate change and improve its agricultural production depends on the environment and natural resources base of the continent, on the impacts of its past and on the nature of its present interactions with the outside world. Africa is a large continent exceeded in land area only by the whole of Asia. In spite of its size, Africa is the least populated of the continents. The human population, which was estimated at nearly 800 million in 2000 (UNEP, 2006), is much smaller than that of India. Africa is usually referred to as dry because the biggest dry land in the world, The Sahara, is in it. The deserts of Africa add up to 1 274 million hectares out of a total land area of more than 3 025 million hectares (UNEP, 2006). But the tropical rainforest area of Africa alone, which receives rain virtually the whole year round, is bigger than India. Therefore, it is not aridity *per se* that prevents Africa from increasing its agricultural productivity and being able to feed all of its people and more.

Africa is the continent that has been the most devastated by slavery, e.g. as recounted by Gray (1961) for Southern Sudan, and colonialism, as recounted by Robinson and Galagher (1965) for the whole continent. And the extraction of able people, both skilled and unskilled, still continues under various forms such as brain drains and economic refuges. No society debilitated thus can be expected to retain its capacity to develop its essential infrastructure, especially for research and development, and build self-sufficiency in agricultural production.

Africa is the most endowed of continents in mineral resources and is rich even in petroleum, though not as rich as the Middle East. These minerals attract much external interference, for example in the eastern Congo (Wikipedia, 2010a). This phenomenon of external perturbation resulting in internal instability is not restricted to the Congo. It is widespread in Africa. Diamonds are among the other minerals that are used to finance conflicts (Wikipedia, 2010b), despite repeated international efforts to control and regulate their trading so that the legitimate governments where these resources are found can use the revenue for the development of their countries. Africa is also well endowed in potential for alternative renewable energy from hydro, geothermal, solar and wind resources.

Therefore, if the mighty outside world would stay outside and allow perturbed Africa to settle down, or, better still, if it would make up for its past perturbing by supporting attempts at stabilization rather than continue to interfere for gain, Africa could develop its agriculture according to its own interests and would be able to feed itself and produce surplus to help feed other parts of the planet.

AGRICULTURAL AND NATURAL BIODIVERSITY IN AFRICA

Africa is often viewed as having a relatively homogenous environment because of the broad vegetation belts – desert, savannah and forest – that stretch from the Atlantic coast on the west to the Red Sea and Indian Ocean coast on the east. This superficial uniformity actually encompasses vegetation types that are complex and rich in biodiversity, comprising a greater diversity of ecosystems than for any other equivalent area of land in the world (White, 1983). White referred to these ecosystem complexes as regional centres of endemism, regional mosaics and regional transition zones (see Figure 1).

Similarly, it has often been assumed that Africa's agriculture was relatively homogenous, and there has been frustration at the lack of improvements in agricultural production from the Green Revolution approach based on the package of 'improved seed' selected to suit a homogenous environment, supported by agrochemicals in the form of fertilizers and pesticides, usually applied under irrigation. For food crops, the focus has been on mainly maize, rice and bread wheat (IAC, 2004). These efforts have ignored sorghum, Africa's most diverse and widely adapted cereal, which has the best ability of all its cereals to provide farmers with genetic resources to mitigate and adapt to climate change (van Oosterhout, 1993), as well as pearl and finger millets and many other crops.

The reality is that diversity is the norm in African farming systems with a farmer typically growing ten or more species and varieties of crops (IAC, 2004) in their home gardens and fields. The traditional farming systems of Ethiopia, for example, use over 100 different crop species (Edwards, 1991). Lack of appreciation of the diversity in both genetic resources and its accompanying traditional knowledge as well as lack of investment in infrastructure and appropriate research and development (R&D) are the main reasons why Africa's agriculture has not developed sufficiently. Other reasons, particularly the ignoring of the role of Africa's women farmers, are well discussed in the Sub-Saharan Africa Report of the International Assessment of Agricultural Knowledge, Science and Technology for Development (IAASTD, 2009).

Dixon *et al.* (2001) carried out a global assessment of farming systems in which they recognized 18 distinct systems in Africa, as shown in Figure 2. The information on these farming systems is summarized in Table 1. The criteria used to differentiate the systems were a) the natural resource base; b) the principal crops and domestic animals; c) the level of crop-livestock interaction; and d) the scale of operations. The agricultural systems are arranged in a descending order of the percentage of the population that they support.

In a study for the International Federation of Organic Agriculture Movements (IFOAM) (Edwards, 2004), the authors found that there was a strong coincidence in the distribution of Africa's floristic regions as identified by White (1983), shown in Figure 1, and the farming systems identified and mapped by Dixon *et al.* (2001), shown in Figure 2.

Four farming systems – maize mixed, cereal/root crop mixed, root crop and agropastoral millet/sorghum – provide the livelihoods for half of the population and occupy 42 percent of the land area in Sub-Saharan Africa. These systems are dominated by smallholder farmers producing and marketing their produce locally. Even within one country, produce rarely moves much outside the agro-ecological zone in which it is grown. One of the biggest challenges for organic producers in Africa who wish to export their products is the lack of infrastructure to get their produce transported and lack of contacts in the importing countries (Taylor, 2010). The situation in North Africa is different because of the easy access to Europe. But

TABLE 1

Farming systems of the African Region with their percentage land area and population, based on Dixon et al. (2001)

FARMING SYSTEM	AGRICULTURAL POPULATION %	LAND AREA %	MAP KEY NUMBER	PRINCIPLE FOOD CROPS AND DOMESTIC ANIMALS
In Sub-Saharan Africa				
1. Maize mixed	15	10	9	Maize, legumes, vegetables, tobacco, cotton, cattle, shoats, chicken
2. Cereal/root crop mixed	15	13	8	Maize, sorghum, various millets, cassava, yams, legumes, vegetables, cattle, chicken
3. Root crop	11	11	7	Yams, cassava, legumes, vegetables, pigs, chicken
4. Agropastoral millet/ sorghum	9	8	11	Sorghum, pearl millet, legumes, sesame, cattle, shoats, chicken
5. Highland perennial	8	1	5	Banana, plantain, enset, coffee, cassava, sweet potato, taro, legumes, cereals, cattle, shoats, chicken, honey bees
6. Pastoral	7	14	12	Cattle, camels, shoats
7. Forest-based	7	11	3	Cassava, maize, legumes, taro, wild fruits and vegetables, cattle, pigs, chicken
8. Highland temperate mixed	7	2	6	Wheat, barley, sorghum, millets (including teff), legumes, oil crops, potato, cattle, shoats, chicken, honey bees
9. Tree crop	6	3	2	Cocoa, coffee, oil palm, rubber, yams, maize, chicken
10. Commercial agriculture small and large holders	5	4	10	Maize, pulses, sunflower, cattle, shoats, chicken, game animals
11. (Coastal) artisanal fishing	2	3	14	(Marine) fish, coconut, cashew nut, banana, yams, fruits, goats, chicken
12. Irrigated (mostly large scale)	2	1	1	Rice, cotton, perennial fruits, vegetables, rainfed crops, cattle
13. Rice/tree crops	2	1	4	Rice, banana, coffee, maize, cassava, legumes, livestock
14. Sparse (arid) agriculture	1	18	13	Irrigated maize, vegetables, date palm, cattle, shoats, chicken
15. Urban-based	3	<1	Not shown	Fruits, vegetables, dairy cattle, shoats, chicken
In North Africa				
16. Highland mixed	30	7	2	Cereals, legumes, sheep
17. Rainfed mixed	18	2	3	Tree crops, cereals, legumes
12. Irrigated	17	2	1	Fruits, vegetables, cash crops
18. Dryland mixed	14	4	4	Cereals, sheep
6. Pastoral	9	23	5	Shoats, barley
15. Urban-based	6	<1	Not shown	Horticulture, chicken
14. Sparse (arid)	5	62	6	Camels, sheep
11. Coastal artisanal fishing	1	1	Not shown	Fish

'Shoats' = mixed flocks of sheep and goats

this agriculture is also dominated by smallholders with nearly half the population concentrated in less than ten percent of the land in the northwest edge of the continent, and along the Nile River in Egypt.

Nearly all the farming systems are mixed, i.e. the farmers keep livestock as well as produce crops. This is because the livestock provide many services, not just meat and milk, for their owners. Without livestock, the essential recycling of nutrients from crop residues and natural vegetation back into the crop land would not take place. This is particularly important for all areas which have marked dry and moist seasons with the farmers dependent on rainfall for their crop production. During the dry season, ruminants break down the plant materials they feed on into manure. This is one of the most important constituents of good quality compost. Until the introduction of chemical fertilizers, the fertility of all types of crop land was maintained through the recycling of animal wastes back to the soil. The majority of Africa's smallholder farmers are still maintaining their animals to help in this recycling of nutrients. If they are properly herded, the other function that domestic animals provide is the maintenance of the vegetation balance in ecosystems. Their hooves can break the crust of dry soil so that rain water can penetrate. They also trample down non-palatable vegetation so that it can break down and get incorporated into the soil, providing good soil structure. Their trampling can also break open the soil so that seeds in the soil seed bank can germinate and the biodiversity in the ecosystem can be maintained (Adams and Butterfield, 2006). These functions of domestic animals are very well understood by pastoralists and agro-pastoralists, as well as to a lesser extent by settled smallholder farmers.

In Sub-Saharan Africa, commercial farming, mostly in Southern Africa, combined with irrigated production in the Niger River basin and along the Nile and its tributaries in the Sudan, engages only seven percent of the population and occupies only five percent of the land. The commercial and irrigated areas are dominated by cash crops and have had, therefore, some investments in infrastructure – dams, irrigation systems, roads and markets – in order to bring the products to national and global markets.

FIGURE 1

The main floristic regions of Africa and Madagascar, from White (1983)

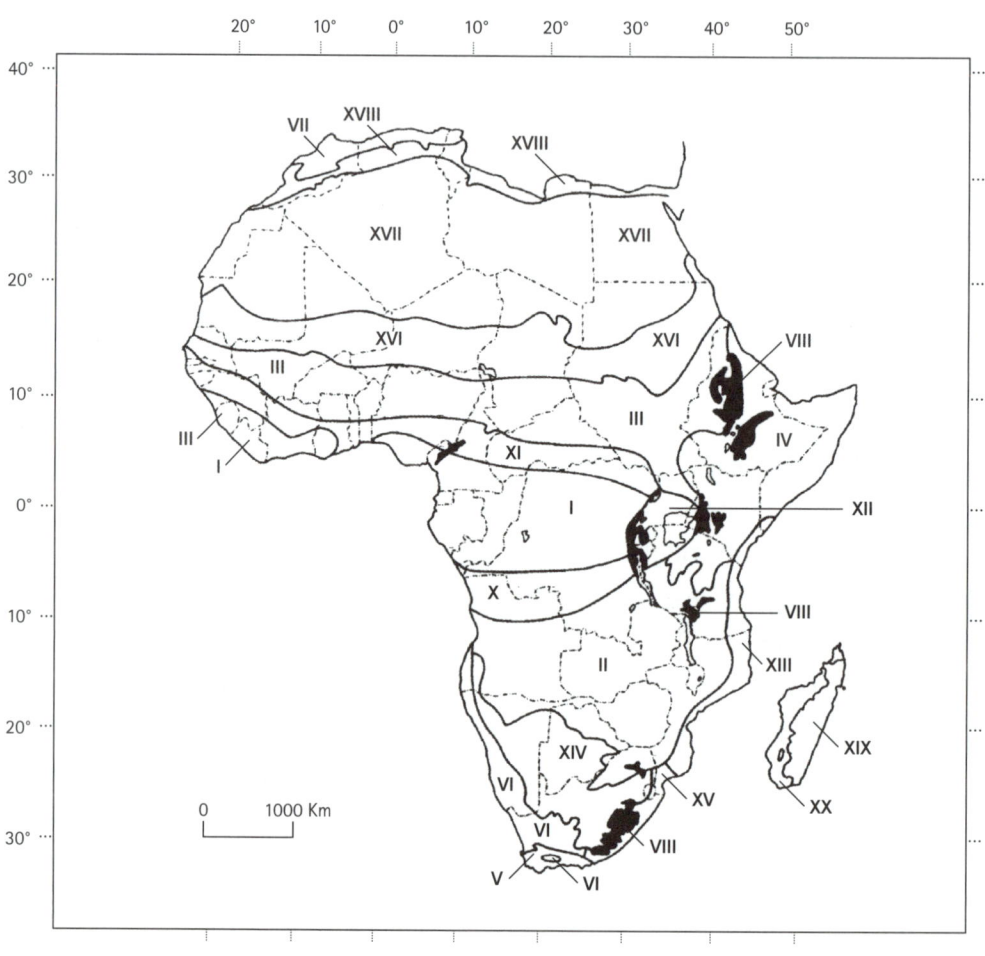

I Guinea-Congolian regional centre of endemism	XI Guinea-Congolian/Sudanian regional transition zone
II Zambezian regional centre of endemism	XII Lake Victoria regional mosaic
III Sudanian regional centre of endemism	XIII Zanzibar-Inhambane regional mosaic
IV Somalia-Masai regional centre of endemism	XIV Kalahari-Highveld regional transition zone
V Cape regional centre of endemism	XV Tongaland-Pondoland regional mosaic
VI Karoo-Namib regional centre of endemism	XVI Sahel regional transition zone
VII Mediterranean regional centre of endemism	XVII Sahara regional transition zone
VIII Afromontane archipelago-like regional centre of endemism, including	XVIII Mediterranean/Sahara regional transition zone
IX Afro-alpine archipelago-like region of extreme floristic impoverishment, where the land goes over 3 000 m	XIX East Malagasy regional centre of endemism
X Guinea-Congolian/Zambezian regional transition zone	XX West Malagasy regional centre of endemism

FIGURE 2

African farming systems, according to Dixon et al. *(2001)*

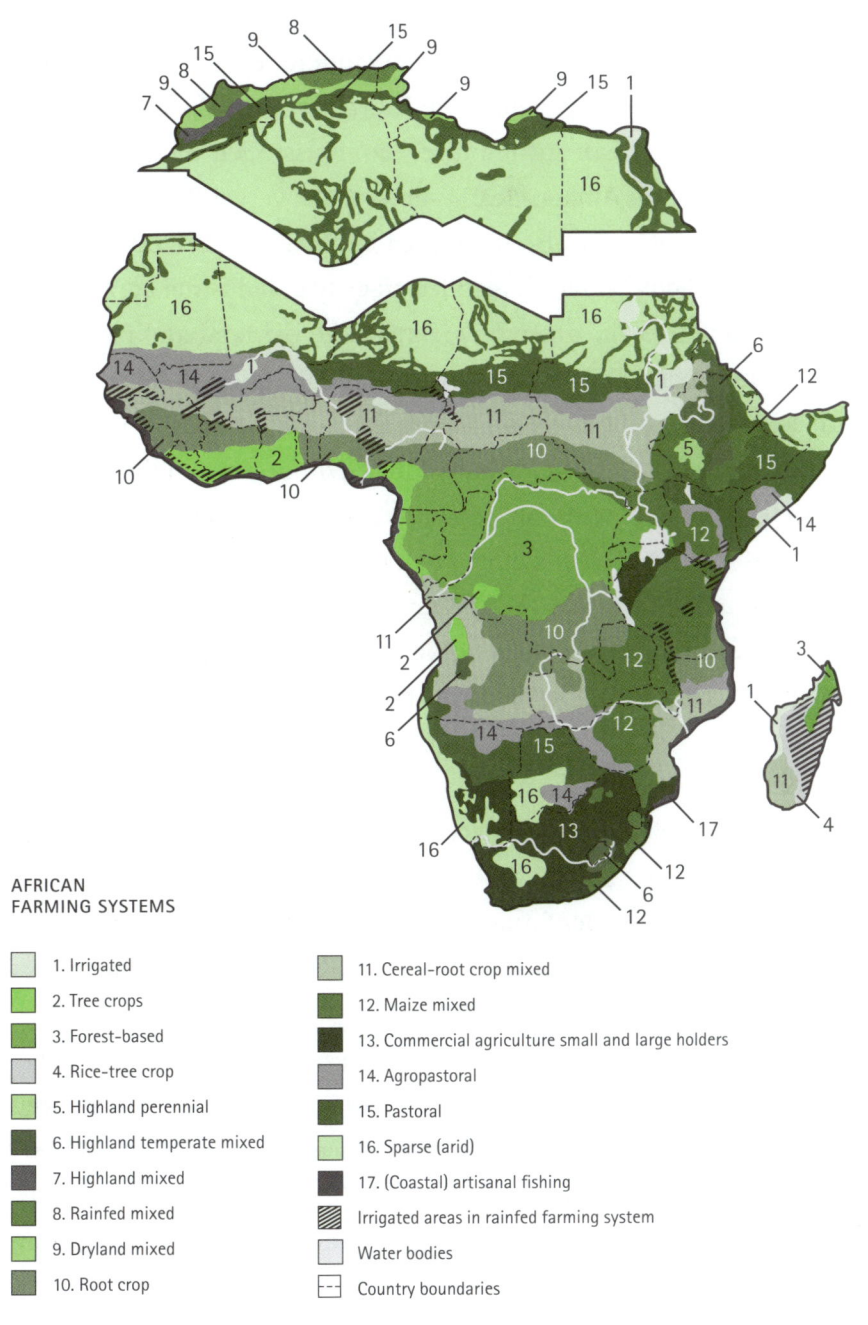

**AFRICAN
FARMING SYSTEMS**

1. Irrigated
2. Tree crops
3. Forest-based
4. Rice-tree crop
5. Highland perennial
6. Highland temperate mixed
7. Highland mixed
8. Rainfed mixed
9. Dryland mixed
10. Root crop

11. Cereal-root crop mixed
12. Maize mixed
13. Commercial agriculture small and large holders
14. Agropastoral
15. Pastoral
16. Sparse (arid)
17. (Coastal) artisanal fishing

Irrigated areas in rainfed farming system

Water bodies

Country boundaries

Overall, however, it is obvious that African smallholder agriculture has not attracted the level of investment needed to raise its productivity through homogenizing and industrializing it, as has been done for industrial agriculture in other continents. Can Africa afford to follow the agricultural development path for industrial agriculture to meet its food requirements? The IAC report (2004), "Realizing the promise and potential of African Agriculture", commissioned by then United Nations Secretary General, Kofi Annan, clearly shows that a very different approach must be used. This position is now also confirmed by the IAASTD (2009).

Kimbrell (2002), in his book "Fatal harvest: The tragedy of industrial agriculture", describes in pictures and text the destructive impacts to the land and farmers of the industrial agriculture model developed by the United States of America, and how these farmers have become more and more dependent on costly solutions. The solutions are costly both in terms of health of the environment and the economy of the farm, as well as the non-renewable resources and personnel needed to run it. The agrochemicals industry, the machinery and mechanics, the highly technical advisors using Geographic Information Systems (GIS) and other data to advise farmers, as well as the immigrant workers needed to provide the labour, make the expenses rise and undermine local food security. Kimbrell also describes and shows the improvements to the environment and human health, including that of the farmers, of growing food organically.

IFOAM issued a press release in February 2010 stating that certified organic produce is now being grown on 35 million hectares of land operated by 1.4 million producers, most of them smallholder farmers (IFOAM, 2010). Although the greatest area of land certified as growing products organically is in Australia, the largest number of certified organic farmers is in Africa.

It is true that when it comes to bulk supply for a population that is largely urbanized and homogenized in its food culture, industrial agriculture has been able to provide the products needed. It has also produced the food for shipments of food aid to people in Africa after they have suffered from drought and conflicts. But this is only because the whole of the industrially managed food chain, from seed through the farm to the market, is heavily subsidized. Without these subsidies, food aid would be unaffordable for developed countries to send to the developing countries, and Africa would be able to develop its own internal markets to provide food from within the continent for its own people, as Zambia was able to do in the food crisis of 2002 when it refused aid food contaminated with genetically modified seed, and opted to buy the food it needed from elsewhere. At the time of the refusal in October 2002, there were over a million metric tonnes of maize available in neighbouring African countries (FAO, 2003). The development of good internal and regional markets in Africa would make it attractive to develop permanent improvements to the infrastructure needed to improve agricultural productivity and transport as well as enhance local R&D to sustain it all. This would greatly benefit the development and expansion of ecological/organic agriculture for both local markets and export.

AFRICA'S CENTRES OF ENDEMISM AND THEIR AGRICULTURE

Most of what is known of global biodiversity is summarized as the number of described species. The Global Biodiversity Assessment of the United Nations Environment Programme (UNEP) used a working figure of 1.75 million described species for the world, which were considered to be only 13 per cent of the 13.6 million species estimated to exist (Heywood and Watson, 1995). The greatest number of species already known is found in tropical rainforests on land and in coral reefs in the sea. The total number of species in a given area generally decreases with its distance from the equator, both north and south, and with increasing altitude. However, the proportion of the known biodiversity is in reverse relationship to its richness with the best known biota being in the temperate and economically developed parts of the world. This state of knowledge for natural biodiversity is equally true for agricultural biodiversity, with the highest number of 'orphan' crops ignored by mainstream research also being found in the tropics and other economically less developed parts of the world (Rehm and Espig, 1991).

For Africa, the flora of higher plants is relatively well known compared to other continents in the tropics (WCMC, 1992; White, 1983). Birds are the best studied and documented group of animals with many identification guides available, for example Sinclair and Ryan (2003) and for invertebrates, butterflies have also been relatively well documented (e.g. Carcasson, 1981). Kingdon has produced a comprehensive compilation of the mammals, published in several parts and summarized in his "Field Guide to African Mammals" (Kingdon, 1997). However, for all other groups of organisms, even the little that is known has only infrequently been compiled and made available.

In identifying the natural biodiversity regions in Africa, Kingdon (1989) uses the fact that around one quarter of the known plants and animals indigenous to Africa are clustered in distinct geographical enclaves referred to as regional centres of endemism. In this, he follows and enriches the work of White (1983). There is a broad correspondence between the distribution of the farming systems and floristic regions in Africa. Therefore, by identifying the farming systems associated with the floristic regions, it is possible to set broad priorities for agricultural intensification based on the organic principles of health, ecology, fairness and caring for the earth. Such intensification could result in improved production for the farmers without extending crop production further into the natural vegetation, and this would benefit the biodiversity of both the farmed and the non-farmed areas.

Following are brief descriptions of the farming systems found in the 11 major African floristic regional centres of endemism and one regional mosaic. The transition zones recognized by White (1983) have not been described though they are shown in Figure 1. This is simply because, as transition zones, their farming systems are mixtures of those of the centres of endemism which are adjacent to them.

Guinea-Congolian regional centre of endemism

This regional centre (I in Figure 1) includes the Congo River basin, which Kingdon (1989) describes as the "huge green belly" of the continent, and the Guinea Forests of West Africa. These are the hottest and wettest parts of the continent.

The natural vegetation inside the Congo River basin is tropical rainforest with swamp forests and edaphic grasslands occurring in areas of impeded drainage. According to White (1983), Fabaceae (Leguminosae) is the family that is the richest in species in this region. The potential for high quality animal feed and for nitrogen rich compost making is thus high.

The main staple crops grown by smallholder farmers are cassava, maize, beans and taro. The people also make much use of wild fruits and vegetables as well as of non-timber forest products, for example honey and caterpillars. The importance of such products to the local economy and to the conservation and sustainable use of forests is now receiving more attention (Crafter *et al.* 1997). Fish are important in the local economy. The number of fish species in the Congo River is more than 400 belonging to 24 families (Lowe-McConnel, 1969). Development of artisanal and improved local fishing along the rivers should learn from the disasters of other areas, notably Lake Victoria, and avoid the introduction of alien species, such as the Nile Perch.

The agriculture of the wet Guinea Forests of West Africa is now dominated by tree crops i.e. cocoa, coffee, oil palm, rubber, though yams and maize are also important (Dixon *et al.*, 2001). Since the mid-1990s, the returns from the four main tree crops, but particularly from coffee and cocoa, have dropped dramatically, and many farmers have become impoverished (Khor, 2010). An ecological/organic approach to intensification and diversification, particularly that which encourages mixed planting of perennial crops for trade with annual crops for local food security, and the use of indigenous legumes for green manure and mulch could help restore the fertility of the land and build resilience for coping with climate change. Where it is being implemented, fair trade combined with organic certification for the marketing of the organic products can bring significantly better returns to the local farmers than normal trade (AdeOluwa, 2010).

Zambezian regional centre of endemism

Much of the Zambezian regional centre (II in Figure 1) is over 900 metres above sea level with rainfall over 1 400 mm in the northern parts, but this decreases to the south and west. The characteristic soils are leached and acidic, often shallow and stony with a hard pan restricting rooting. White (1983) identified over half of the plant species occurring in this region as endemic. The most widespread and characteristic vegetation type of the Zambezian regional centre of endemism is woodland. Tree branches are regularly lopped and burnt to improve the fertility of the soil, a system called "citimene". This has had a marked impact on the dynamics of the vegetation with the trees being more or less uniform in appearance.

Four agricultural systems are found in this regional centre of endemism: root crops and highland perennial crops in the wetter areas to the north, and mixed maize and other cereals and root crops in the drier south and west respectively. Low soil fertility is a major constraint to raising crop production. The promotion of animal production through a holistic management of communal herds (Adams and Butterfield, 2006) and composting would rehabilitate the environment, as well as improve soil fertility and human nutrition.

Sudanian regional centre of endemism

The Sudanian regional centre of endemism (III in Figure 1) forms a belt from the west coast in Guinea to the Red Sea coast in Sudan and Eritrea. The belt widens across the middle and reaches into northern Uganda, and occupies most of Southern Sudan as well as the western lowlands of Ethiopia. The land lies mostly between 500 and 700 metres above sea level. The rain falls between May/June and September giving around 1 000 mm per annum in the south, where there is woodland, to about 600 mm per annum in the grasslands of the north. This area also includes important wetlands: the diminishing Lake Chad, the Niger River delta in the west, and the Sudd and White Nile in the east. Fires are a regular feature of the slash and burn type of agriculture that takes place because the tall grasses become silicified and inedible for animals. The fires clear the way for new plant growth for both domestic and wild animals.

The biodiversity includes several trees of economic importance, e.g. *Acacia senegal*, which produces gum arabic, *Balanites aegyptiaca* (the desert date), which has edible fruits and produces an edible oil, and the endemic *Butyrospermum paradoxum*, the shea butter tree.

Cereal/root crop mixed farming is the dominant agricultural system of this region. It is considered that sorghum (*Sorghum bicolour*) cultivation evolved in this region, along with cowpea (*Vigna unguiculata*), sesame (*Sesamamum indicum*), watermelon (*Citrullus lanatus*) and roselle (*Hibiscus sabdariffa*) (Harlan *et al*, 1976). Pastoralism is also important, but greater attention could be given to improving productivity along with environmental rehabilitation. With mixed farming and composting, agricultural productivity can easily be raised.

Somalia-Masai regional centre of endemism

The Somalia-Masai regional centre of endemism (IV in Figure 1) is dry, with rainfall rarely exceeding 500 mm a year. Over half of the 2 500 plant species found are endemic to this regional centre of endemism. Most of the vegetation is deciduous bushland and thickets that give way to semi-evergreen and evergreen bushland on the lower slopes of mountains. Seasonally waterlogged areas become grasslands. This floristic region is the source of incense and myrrh, which are gums collected from species of *Boswellia* and *Commiphora* respectively. Other species with high traditional and potential wider economic value include the Yehib nut, *Cordeauxia edulis*.

Agropastoralism and pastoralism have been practiced throughout this floristic region for a very long time, probably as long as cultivated agriculture in the Ethiopian highlands, i.e. for at least 5 000 years. The people, the vegetation and the wildlife are hardy and have co-evolved to cope with a very harsh and variable climate. This has been well documented for Masai pastoralism (FAO, 2003).

Traditional crop cultivation is important along the edges of rivers. This can be expanded through water harvesting and irrigation using ground water. The existing animal production would make it easy to make compost of high quality and maintain high productivity in the irrigated agricultural areas.

Cape regional centre of endemism

The Cape region (V in Figure 1) is floristically the richest part of Africa. There are over 7 000 species in this very small area, of which more than half are endemic (White, 1983). The plants are adapted to cool moist winters contrasting with hot and dry summers with frequent fast and fierce summer fires. Many of the species in this centre of endemism have been taken into cultivation and developed into important horticultural plants, for example, the Proteas, Heaths and Heathers, as well as many bulbous plants. Rooibos and honeybush tea have been developed from local species and now have an international market as speciality teas, including with organic certification. Modern commercial agriculture based on crops from the Mediterranean region, particularly grapes, has expanded into this area. A shift to composting for maintaining high soil fertility would be easy because animal production is already important, though it could be expanded to all farms.

Karoo-Namib regional centre of endemism

The Karoo-Namib regional centre of endemism (VI in Figure 1) is a strip of arid land stretching from the Cape for 2 000 kilometres up the western Atlantic Ocean coast through Namibia to Angola, and stretching inland to the Orange River. Half of the 4 000 species of plants found in the area are endemic to it. All the species are adapted to withstanding long periods of drought and, when moisture comes, to responding very fast in growth and flowering. Many of the annuals are grown as horticultural varieties to decorate public parks and gardens because of their very hardy nature.

The dryness mostly restricts agriculture to transhumant pastoralism and to extensive ranches of low carrying capacity. However, even in this region, there is some crop cultivation, especially where some supplementary irrigation is possible.

Mediterranean regional centre of endemism

This area (VII in Figure 1) is dominated by the Atlas Mountains and is found at the north-western edge of the continent. Around 4 000 species occur in this area, but only 20 percent of them are endemic to North Africa.

There are three agricultural systems found in the Atlas Mountains and along the area between the mountains and the Mediterranean Sea. These are the highland mixed, rainfed mixed, and dryland mixed farming systems. Over 60 percent of the agricultural population obtain their livelihoods from these three systems. Since the farming systems are already mixed and there is a large agricultural labour force, i.e. producing crops and animals, high quality composting for intensive ecological crop and animal production can be easily introduced throughout.

Afromontane archipelago-like regional centre of endemism

The Afromontane region (VIII in Figure 1) is scattered on the mountains and highlands which are found in every country in Eastern and Central Africa, as well as in Cameroon in West Africa and the mountains in Southeast Africa between the Republic of South Africa and Mozambique. All these highland areas are generally wetter than the surrounding lowlands, and are thus important water towers for these areas.

Taken as a whole, the Afromontane floristic region has about 4 000 species of which about 75 percent are endemic (White, 1983) with about one-fifth of the genera also being endemic. The vegetation is forest and woodland adapted to alternating dry and wet periods on the sloping areas up to around 3 000 metres above sea level with edaphic grasslands and wetlands on the flatter areas.

The main agricultural types in this regional centre of endemism are highland perennial and highland temperate mixed farming systems. Both systems combine animal husbandry with crop cultivation and grow a wide range of crop species and varieties within the species. The greatest agricultural biodiversity is found in Ethiopia, which has been recognized as one of the world's Vavilov centres for crop genetic diversity. There is little flat land in the Afromontane region. Therefore, soil and water conservation are critical for the survival of the farmers, and many ingenious indigenous systems have been developed (Reij *et al.*, 1996). Since the farming systems are mixed, high quality compost can easily be made by the farmers, as has already been shown in Ethiopia (Edwards *et al.*, 2007).

Afro-alpine archipelago-like regional centre of endemism

White (1983) refers to the Afro-alpine areas of Africa (IX, not shown separately in Figure 1) as "a region of extreme floristic impoverishment". In the sense that the species numbers in the flora are low, he is right. But Hedberg (1969) has shown that 80 percent of the plant species in the Eastern Afro-alpine region are endemic to that region.

The environment is subjected to extremes of temperature every day. It freezes at night, and day temperatures become hot, especially when the skies are clear. The plants are thus rooted in cold soil, but their aerial parts can have high temperatures. This condition gives rise to a range of adaptations, e.g. cushion forms, acaulescence, succulence, and plants which close up tightly at night and open up during the day. Plant growth is slow. Therefore, though the environment is wet the whole year round, the vegetation is vulnerable because it cannot recover quickly after it has been disturbed, let alone destroyed.

The Afro-alpine areas are important as water towers for their surrounding lowers areas. In Ethiopia, people live in these areas cultivating barley and grazing their sheep and cattle. In many other parts of Africa, for example on Mounts Kenya and Kilimanjaro, the most important economic activity is tourism.

Zanzibar-Inhambane regional mosaic

This area (XIII in Figure 1) has been known as the spice coast for more than 2 000 years. It includes the islands of Zanzibar and Pemba. The distinctive vegetation occupies a narrow strip along the Indian Ocean coast starting from the southern tip of Somalia and continuing south into Mozambique. Due to the influence of the warm Indian Ocean, this area has been relatively unaffected by major climatic changes for about 30 million years, and this has resulted in high diversity and endemism, particularly in the Usambara Mountains.

Of the 190 tree species found in this regional mosaic, 92 are endemic. The non-timber forest flora is also rich with one of the best known groups being the African violets, *Saintpaulia* in the family Gesneriaceae, which has about 20 indigenous species. These species probably evolved from seed blown in from Madagascar, with each species developing to be adapted to a relatively small area of the Usambara Mountains. In cultivation, these plants hybridize readily so that there are now thousands of cultivars supporting a world trade estimated at over 30 million dollars annually in the 1980s (Kingdon, 1989).

The economically important products include cardamom, cloves, nutmeg, pepper, vanilla, coconut and cashew nut. Apart from felling trees for timber, the clearing of forest undergrowth to increase these spice plantations is one of the major threats to the biodiversity of this area. Many food crops are, and can be, grown. Animal production should be given a greater focus than it has been receiving. Composting could be introduced easily, especially with the further development of mixed farming.

East Malagasy regional centre of endemism

Eastern Madagascar (XIX in Figure 1) is dominated by the central highlands with mountains up to and above 2 000 metres above sea level. To the west, this floristic region descends to about 800 metres above sea level. To the east, the central highlands end abruptly in steep escarpments overlooking the narrow coastal plain. The area is wet with annual rainfall exceeding 3 000 mm a year in some places. Rainfall decreases towards the centre of Madagascar, which is both drier and colder. Extensive marshes and lagoons are found in the coastal plains.

There are about 6 100 species in the floristic region, with 4 800 of them (about 80 percent) being endemic. Of the 1 000 genera, 160 (16 percent) are endemic.

The vegetation of the lowlands is rainforest, while higher up there is a mosaic of moist montane forest and drought-resistant montane forest. When deforested, the leached porous soils are covered by bamboo thickets. Compact ferralitic soils, however, are covered by grasslands following deforestation. Fire then keeps them permanently so. Therefore, the vegetation is highly vulnerable. Though there still are extensive areas of forest, deforestation is going on fast.

The main agricultural system is dominated by rice production. An emphasis on agroforestry especially along the edges of the rice fields can be used as the basis for composting and raising soil fertility. The expansion of the system of rice intensification (SRI) that was developed over 25 years ago by Fr. Henri de Laulaniè working with local farmers, also offers the possibility of improving crop production without extending cultivation into more areas of the fragile forest ecosystems. This management system has enabled poor farmers to raise their rice yields from an average of two to eight tonnes per hectare (Uphoff, undated).

West Malagasy regional centre of endemism

The West Malagasy regional centre of endemism (XX in Figure 1) is found on the flat plains inside the western coastline. The driest parts are in the south where rainfall can be as little as 300 mm per annum. The central plains generally receive about 500 mm, but this increases up to 2 000 mm per annum in the northwest.

There are about 2 400 plant species, of which about 1 900 (79 percent) are endemic. Of the about 700 genera, some 140 (20 percent) are endemic.

The vegetation varies from dry deciduous forests to deciduous thicket and grasslands. These grasslands are extensive, covering about 80 percent of the area. They are secondary in origin, having been caused by deforestation and being maintained by regular fires. Mixed agriculture is possible in most parts. Therefore, composting can be used to raise soil fertility.

HOW CAN AGRICULTURAL PRODUCTION BE INTENSIFIED IN AFRICA?

This is a frightening time of climate change. The intensification of agricultural production in Africa based on ecological principles can take place without the use of industrially produced agrochemicals, especially fertilizers that are made from fossil fuels. Experience in Ethiopia (Edwards *et al.*, 2007; Edwards *et al.*, 2010) has shown this to be possible. Preparing compost from household and farm waste and using it to raise soil fertility has been found to be as effective as, and in the case of crops bred by smallholder farmers, to be more effective than, using chemical fertilizers to raise agricultural productivity. The role that the holistic management of domestic livestock in this recycling process and in maintaining ecosystem services needs to be much better understood and supported by professional agriculturists as well as policy-makers.

There is also the possibility of extending the improved agronomic management of crops based on the system of rice intensification (SRI) developed in Madagascar[1] (Uphoff, undated) and now spreading rapidly throughout much of south west Asia, for example in India (Anonymous, 2010). The World Bank is also supporting the SRI initiative (World Bank Institute, 2008).

But, above all, peace must prevail in Africa if the needed intensification of agricultural production is to be achieved. As Devereux (2001) pointed out, "virtually every [African] country that has suffered famine in the past twenty years has suffered a war at the same time".

---

1   Other crops such as finger millet, wheat, sorghum and faba bean can also be grown using the management practices described for rice. This can be called the 'System of Crop Intensification' (SCI).

EXISTING AGRICULTURAL POLICIES IN AFRICA

Following are some generalizations based on the international laws to which African countries are parties and on the documents of the African Union including those of the New Partnership for Africa's Development (NEPAD).

Article 8(j) of the Convention on Biological Diversity (CBD) recognizes the rights of local and indigenous communities. Based on this, the Organization of African Unity, which became the African Union (AU), approved its "Model Law on the Rights of Local Communities, Farmers and Breeders, and on Access to Biological Resources" in Ouagadougou in 1998 (Ekpere, 2001). Article 9 of the International Treaty on Plant Genetic Resources for Food and Agriculture gives international recognition to the right of a country to recognize farmers' rights through its domestic law.

The section of NEPAD that deals with agriculture (NEPAD, 2001) recognizes the problems of agricultural intensification in Africa as being "biases in economic policy and instability in commodity prices" as well as climate uncertainty. It sees the solution as lying in increasing investment, in "the encouragement of local community leadership in rural areas and [in] the involvement of these communities in policy [making] and the provision of services". The African Union and NEPAD subsequently translated these principles into a detailed agricultural development programme (AU and NEPAD, 2003). Recalling the known fact that most African farmers are women, of particular relevance is the emphasis given by this agricultural development programme to women in rural development. The facts on the ground in Africa, the relevant African Union and NEPAD documents and the relevant international laws thus reinforce one another to focus on communities of smallholder farmers as the main force for intensifying agricultural production in Africa.

CONCLUSION

Generally the larger and more conspicuous plants, for example flowering plants and particularly trees, and animals are relatively well known both globally and regionally (WCMC, 1992). But much less is known of most of the 'lower' and smaller organisms, and hardly anything about the micro-organisms, except for those that cause disease or are classified as pests. This general situation is also true for Africa. However, individuals and organizations from outside Africa are collecting and exploiting this diversity with little to no involvement of African scientists and local institutions. It is the smaller organisms and micro-organisms that contribute much to the overall health of ecosystems, and particularly to the essential recycling of carbon and other nutrients (WBGU, 2001). Therefore, any system of agriculture based on ecological/organic principles will work with the natural cycles rather than against them and has the inherent potential to be sustainable. This principle underlies the inbuilt robustness, or resilience, of Africa's traditional agricultural systems.

In conclusion, the fact that African agriculture is primarily subsistence and is carried out by smallholder farmers, that most of these smallholder farmers are women, that climate chaos is a real global threat to agriculture, and that food production has thus to adapt to the droughts, heavy rains and floods exacerbated by it, coupled with the fact that the subsistence farming systems in Africa are still more or less intact and thus versatile, dictate that intensification of agricultural production in Africa be ecological and that agricultural systems remain as diverse as the ecological diversity of the continent requires. The needed intensification might thus sound frighteningly complex. But, it is not if it remains centred on local farming communities who are, themselves, already as diverse in their agricultural lore as their respective ecosystems and thus need only relatively small inputs of scientific information that is new to them.

REFERENCES

Adams, A. & J. Butterfield. 2006. The essence of holistic management. *In: Practice.* Special Edition 2006. Holistic Management International & Africa Center for Holistic Management, Victoria Falls, Zimbabwe.

AdeOluwa, O.O. 2010. Organic agriculture and fair trade in West Africa. *In: Climate change and food systems resilience in Sub-Saharan Africa.* FAO, Rome. (Accepted for publication).

Anonymous. 2010. State promoting SRI cultivation, says Rout. *The Hindu.* 21 March 2010. Available at http://www.hindu.com/2010/03/21/stories/2010032151750300.htm (accessed 25 March 2010; verified 13 April 2010).

AU & NEPAD. 2003. *Comprehensive Africa Agriculture Development Programme.* Midrand, South Africa.

Carcasson, R.H. 1981. *Collins handguide to the butterflies of Africa.* William Collins Sons & Co. Ltd., London.

Crafter, S.A., J. Awimbo & A.J. Broekhoven (eds.). 1997. *Non-timber forest products – Value, use and management issues in Africa, including examples from Latin America.* IUCN, Gland, Switzerland.

Devereux, S. 2001. Famine in Africa. *In* S. Devereux & S. Maxwell (eds.) *Food security in Sub-Saharan Africa.* Intermediate Technology Development Group, London.

Dixon, J., A. Gulliver & D. Gibbon. 2001. *Farming systems and poverty: Improving farmers livelihoods in a changing world.* FAO, Rome & World Bank, Washington DC.

Edwards, S. 1991. Crops with wild relatives found in Ethiopia. *In* J.M.M. Engels, J.G. Hawkes & Melaku Worede (eds.) *Plant genetic resources of Ethiopia.* Cambridge University Press, Cambridge, UK.

Edwards, S. 2004. The biodiversity of Africa, and the contribution of organic agriculture to Africa's development. Study undertaken for IFOAM.

Edwards, S., Arefayne Asmelash, Hailu Araya & Tewolde Berhan Gebre Egziabher. 2007. *Impact of compost use on crop yields in Tigray, Ethiopia, 2000-2006 inclusive.* FAO, Rome. Available at http://www.fao.org/documents/pub_dett.asp?lang=en&pub_id=237605 (verified 13 April 2010).

Edwards, S., Tewolde Berhan Gebre Egziabher & Hailu Araya. 2010. Successes and challenges in ecological agriculture: Experiences from Tigray, Ethiopia. *In: Climate change and food systems resilience in Sub-Saharan Africa.* FAO, Rome. (Accepted for publication).

Ekpere, J.A. 2001. *The African Model Law [for] the Protection of the Rights of Local Communities, Farmers and Breeders, and for the Regulation of Access to Biological Resources.* Organization of African Unity (OAU), Scientific, Technical and Research Commission of the OAU, Gaia Foundation and Institute for Sustainable Development, Addis Ababa.

FAO. 2003. Case Study No. 12: Globally Important Ingenious Agricultural Heritage Systems. *In*: *Biodiversity and the ecosystem approach in agriculture, forestry and fisheries.* Proceedings of the satellite event on the occasion of the Ninth Regular Session of the Commission on Genetic Resources for Food and Agriculture: Rome, 12-13 October 2002. FAO, Rome.

Gray, R. 1961. *A history of the Southern Sudan, 1839-1889.* Oxford University Press, London.

Harlan, J.R., J.M.J De Wet & A.B.L. Stemler. 1976. Plant domestication and indigenous African agriculture. *In* J.R. Harlan, J.M.J De Wet, & A.B.L. Stemler (eds.). *Origins of African plant domestication.* Mouton Publishers, The Hague.

Hedberg, O. 1969. Evolution and speciation in a tropical high mountain flora. *In* R.H. Lowe-McConnel (ed.) *Speciation in tropical environments.* Academic Press, London.

Heywood, V.H. & R.T. Watson (eds). 1995. *Global Biodiversity Assessment.* UNEP and the University Press, Cambridge, UK.

IAASTD. 2009. *Agriculture at a crossroads.* International Assessment of Agricultural Knowledge, Science and Technology for Development. Island Press, Washington DC.

IAC. 2004. *Realizing the promise and potential of African agriculture.* InterAcademy Council, The Netherlands.

IFOAM. 2010. The world of organic agriculture – Statistics and emerging trends. Press Release 15 February 2010. Available at http://www.ifoam.org/press/press/2008/statsbook2010.php (accessed 22 March 2010; verified 13 April 2010).

Khor, M. 2010. Trade policy implications for Africa's agriculture. *In: Climate change and food systems resilience in Sub-Saharan Africa.* FAO, Rome. (Accepted for publication).

Kimbrell, A. 2002. *Fatal harvest: The tragedy of industrial agriculture.* Island Press, California, USA.

Kingdon, J. 1989. *Island Africa: The evolution of Africa's rare animals and plants.* Princeton University Press, USA.

Kingdon, J. 1997. *Kingdon field guide to African mammals.* Academic Press, USA.

Lowe-McConnel, R.H. 1969. Speciation in tropical freshwater fishes. *In* R.H. Lowe-McConnel (ed.) *Speciation in tropical environments.* Academic Press, London.

NEPAD. 2001. *The New Partnership for Africa's Development.* Abuja, Nigeria.

Rehm, S. & G. Espig. 1991. *The cultivated plants of the tropics and subtropics: Cultivation, economic value, utilization* (original in German as *Die Kulturpflanzen der tropen und subtropen*, translated by G.M. McNamarra & C. Ernsting). Verlag Josef Margraf, Weikersheim.

Reij, C., I. Scoones & C. Toulmin. 1996. *Sustaining the soil: Indigenous soil and water conservation in Africa.* Earthscan, London.

Robinson, R. & J. Gallagher. 1965. *Africa and the Victorians: The official mind of imperialism.* Macmillan & Co Ltd., London.

Sinclair, I. & P. Ryan. 2003. *Birds of Africa south of the Sahara.* Struik Publishers, Cape Town, South Africa.

Taylor. A. 2010. Organic trade promotion in Uganda: A case study of the EPOPA project. *In: Climate change and food systems resilience in Sub-Saharan Africa.* FAO, Rome. (Accepted for publication).

Uphoff, N. (undated) Responses to frequently-asked questions about the system of rice intensification (SRI). Cornell International Institute for Food, Agriculture, and Development (CIIFAD), Cornell University, Ithaca, New York. Available at http://www.wassan.org/sri/SRI_Responses/SRI_Responses.htm (accessed 25 March 2010; verified 13 April 2010).

UNEP. 2006. *Africa Environment Outlook 2.* UNEP, Nairobi, Kenya.

van Oosterhout, S. 1993. Sorghum genetic resources of small-scale farmers in Zimbabwe. *In* W. de Boef, K. Amanor, K. Wellard with A. Bebbington (eds.) *Cultivating knowledge: Genetic diversity, farmer experimentation and crop research.* Intermediate Technology Publications, London.

WBGU (Wissenschaftlicher Beirat der Bundesregierung Globale Umweltveränderungen – German Advisory Council on Global Change). 2001. *World in transition: Conservation and sustainable use of the biosphere.* Earthscan, London.

WCMC (World Conservation Monitoring Centre). 1992. *Global biodiversity: Status of the earth's living resources.* Chapman & Hall, London.

White, F. 1983. *The Vegetation of Africa: A descriptive memoir to accompany the Unesco/AETFAT/UNSO vegetation map of Africa.* UNESCO, Paris.

Wikipidea. 2010a. United Nations Mission in the Democratic Republic of Congo. Available at http://en.wikipedia.org/wiki/United_Nations_Mission_in_the_Democratic_Republic_of_Congo (last modified 30 January 2010; accessed 22 March 2010; verified 13 April 2010)

Wikipedia. 2010b. Blood diamond. Available at http://en.wikipedia.org/wiki/Blood_diamond (last modified 2 March 2010; accessed 22 March 2010; verified 13 April 2010).

World Bank Institute. 2008. System of Rice Intensification: Achieving more with less – A new way of rice cultivation. Available at http://info.worldbank.org/etools/docs/library/245848/index.html (accessed 25 March 2010; verified 13 April 2010).

THE GREEN REVOLUTION IN ASIA: LESSONS FOR AFRICA

Hira Jhamtani

CONTENTS

TABLES

ABOUT THE AUTHOR

Hira Jhamtani is a researcher on environment and development issues in Asia and Indonesia. She was a board member of the Indonesian Society for Social Transformation (INSIST) and is an associate of the Third World Network.

INTRODUCTION

Asia, the rice barn of the world, has often been cited as a "succesful" continent in implementing the Green Revolution, particularly in rice production. Some experts and bureaucrats have said that without the Green Revolution in the 1960s and 1970s, "Asia would have suffered from famine". How true this is remains a question.

In any case, yield increases, while certainly important, do not go to the heart of the problem. The real challenge facing the developing world is achieving food security at the national and household levels. The two tables below show that it is the food security situation that needs further scrutiny.

Table 1 shows that there are about 500 million people in Asia (and the Pacific) still hungry. What is even more worrying is that most of the hungry people are in fact from farm households (400 million as in Table 2), who are supposedly the food producers. The fact that farm households are the most hungry must be a point of critical reflection about the current agriculture system.

TABLE 1

Where are the hungry?

REGION	TOTAL (MILLIONS)
India	214
Sub-Saharan Africa	198
Asia & Pacific	156
South America	56
China	135

Source: FAO (2003 from www.developmenteducation.ie)

TABLE 2

Who are the hungry?

REGION	TOTAL (MILLIONS)
Farm Households	400
Rural Landless	160
Urban Households	64
Herders, Fishers and Forest Dependents	56

Source: UN Hunger Task Force (2003 from www.developmenteducation.ie)

The Green Revolution package of high-yielding varieties (HYVs), irrigation and agrochemicals is often seen as mainly a technological intervention to boost food production. However, as experience has shown, the Green Revolution is also a socio-economic and political construct. It also has an environmental dimension as agriculture is mainly based on natural resources.

Many valuable lessons can be drawn from the Green Revolution experience in Asia. Those lessons can hopefully help Africa to make strategic policy decisions before embarking on an agricultural revolution of any kind.

In Asia, a growing number of farmers are turning back to non-chemical or less-chemical agriculture as the costs of the Green Revolution system have risen while yields have stagnated. This in itself is an issue that must be considered carefully for Africa. There is also a whole range of information and knowledge now on alternative systems of agriculture that are adapted from traditional local systems, as well as based on empirical and conventional scientific studies.

PRODUCTION INCREASE NOT SUSTAINABLE

The main achievement of the Green Revolution in Asia has been the increase in grain production. Taking the example of rice, production did increase, and many experts attributed that to HYV seeds and chemical inputs only. However, other factors were important such as subsidies for the inputs and development of new varieties, access to credit and markets, irrigation systems, government price support and infrastructure improvement (mainly roads and transport systems). What is also important is that the single rice crop system was changed to two or three monocroppings of rice, which also explained the increase in overall production. Area expansion contributed to nearly one-third of Asian rice output growth in the 1960s and one-fifth in the 1970s (Pingali and Rosegrant, 1994).

The increase in production is, however, not sustainable over a long period. As early as 1976, a report from the Asian Development Bank (ADB) noted that the growth rate in rice yields between 1963–67 and 1971–75 was less than 1.5 percent per annum for South and Southeast Asia as a whole and below one percent for several countries. And this was for irrigated (or wet) rice fields; there was no evidence of a major breakthrough in dry land agriculture (Morgan, 1978). Another study reported that rice yield growth in Asia declined sharply in the 1980s, from an annual growth rate of 2.6 percent in the 1970s to 1.5 percent during the period beginning in 1981 (Pingali and Rosegrant, 1994). Disregarding the discrepancy in figures, the conclusion is that the increase in yields is not sustainable.

An interesting example is Indonesia. Real increase in production of up to 3.52 percent per annum was achieved only for ten years during 1979–1989. Since then, total rice production growth has declined to 1.04 percent per year with a productivity increase of only 0.05 percent per year (Swastika *et al.*, 2002, cited in Jhamtani, 2008).

Indonesia was acclaimed by the Food and Agriculture Organization of the United Nations (FAO) for attaining rice self-sufficiency in the 1980s but the country has had to import rice again since 1994. Indeed, farm-level evidence from the rice bowls of Asia indicates that intensive rice monoculture systems (as implemented through the Green Revolution) led to declining productivities of inputs over the long term (Pingali *et al.*, 1997, cited in Lim, 2009).

The decline in yields, but increase in the price of inputs, has also had an adverse impact on the economic welfare of rural communities. This has led to an urban drift and the creation of increasing numbers of urban poor. The Indonesian experience with the Green Revolution indicates that it was focused on increasing rice production rather than farmers' income and that the programme was not cost-efficient and required huge funding (Pribadi, 2001, cited in Jhamtani, 2008). Africa needs to take note of this. The issue is how to sustain productivity as well as farmers' income, in a manner that is cost-efficient.

GREEN REVOLUTION LIMITS

One of the most important features of the Green Revolution is the use of agrochemicals (fertilizers, pesticides, herbicides), a feature that was non-existent before then. Agrochemicals are used because High-yielding varieties (HYV) were constructed to be responsive to chemical fertilizers and were more susceptible to pest outbreaks. In developing countries, these chemicals are costly. And, over a few years, more chemicals have to be used to achieve the same yield. In Indonesia, the application of fertilizers in rice production between 1975 and 1990 rose from under 25 kg/ha to over 150 kg/ha (Fitzgerald-Moore and Parai, 1996, unpublished).

Thus, the increase in yields is offset by the increase in cost associated with increased use of chemicals. In the Central Plains of Thailand, yield increased 6.5 percent but fertilizer use increased by 24 percent and pesticides by 53 percent. In West Java, Indonesia the 23 percent increase in yield was virtually cancelled out by 65 percent and 69 percent increases in fertilizers and pesticides respectively (Rosset *et al.*, 2000, cited in Lim, 2009).

Agrochemicals not only increase production costs but also have health and environmental impacts, whose costs have not been properly internalised in the calculation of yields and production costs under the Green Revolution system. For example, the costs associated with agrochemical pollution of water systems and soils have never been taken into account. Accidents, and even deaths, of farmers and agricultural labourers due to lack of knowledge on safe use of chemicals have also been underreported. A study showed that poisoning episodes occur largely during spraying, mixing, and diluting of pesticides or due to the use of malfunctioning or defective equipment among agricultural workers (Jeyaratnam, 1993). Most farmers are not well educated on this and not enough information has been given to them on the safe handling of the chemicals.

Easy access to pesticides has also meant that these chemicals have become a common means of committing suicide among farmers when they go bankrupt or are embroiled in debts that they cannot repay. In many cases debts were incurred when they borrowed money to buy expensive inputs such as seeds and agrochemicals; when the harvest failed or prices dropped, they could not pay back their debts. These are externalities that can cancel out the benefits arising from increased yields.

At the consumer level, pesticides have contaminated food, leading to health problems. In Indonesia, Dicloro-Difenil-Tricloroetan (DDT) residue was found even in mother's milk (Buchory, 1999, cited in Jhamtani, 2008). The price paid for chemical contamination thus goes far beyond the agriculture fields to our daily diet.

It is interesting to note that since at least the year 2000, the International Rice Research Institute (IRRI) has recognised that high dependence on agrochemicals may not be the solution to the challenges confronting agriculture. In a press release in July 2004, IRRI said that 2 000 poor rice farmers in Bangladesh have proven over the course of two years – four seasons – that insecticides are a complete waste of time and money. When they stopped spraying, yields did not drop – and this was across 600 fields in two different districts over four seasons.

This was a finding from the Livelihood Improvement Through Ecology (LITE) project, a joint project between IRRI and the UK Department for International Development (DfID), which has demonstrated that insecticides can be eliminated and nitrogen fertilizer (urea) applications reduced without lowering yields. Similar studies in the Central Luzon province of the Philippines and in certain parts of Vietnam have already demonstrated that pesticides were not required (Sharma, 2004). Again, this is a valuable lesson for Africa as it considers whether to embark on chemical-intensive agriculture.

In a similar manner, the use of HYVs is also not sustainable mainly due to their characteristic of being genetically uniform. The FAO has warned of a large-scale loss of plant genetic diversity and the erosion of biodiversity. This happens at two levels. First, genetic diversity is reduced when monocultures of rice and wheat replace mixtures and rotations of diverse crops such as wheat, maize, millets, pulses, and oil seeds. Secondly, genetic diversity is reduced because the HYVs of rice and wheat come from a narrow genetic base. As the genetic background of such crops is narrow, their ability to resist diseases and pests has declined relative to the ability of diseases and pests to overcome the resistant traits that have been bred into the seeds (Fitzgerald-Moore and Parai, 1996, unpublished). Each new HYV that has become susceptible to pests and diseases has to be replaced, which involves costs.

Although HYVs are bred to resist insects, diseases and environmental stresses, when planted over a large area, they are in fact more vulnerable to pests. When single cultivars, such as the IR-36 rice, cover large numbers of fields, infestation can spread like wildfire, as was the case in Indonesia. By 1977, 1.5 million ha of rice fields in the main growing areas of Java, Sumatra and Bali islands were infested by the brown hopper, followed by the tungro virus a few years later. The pests had become resistant to pesticides by then. Replacing IR-36 with the IR-64 variety did not solve the problem as IR-64 became susceptible to stem borer after a few years (Oka, 1995, cited in Jhamtani, 2008). Indonesia also lost about 1 000 local rice varieties when it implemented the Green Revolution system.

The lesson to be learnt is that intensive double or triple monocropping of rice has caused degradation of the paddy micro-environment and reductions in rice yield growth in many irrigated areas in Asia. Problems include increased pest infestation, mining of soil micronutrients, reductions in nutrient-carrying capacity of the soil, build-up of soil toxicity, and salinity and waterlogging (Pingali and Rosegrant, 1994). Thus, areas that have not experienced intensification and the Green Revolution system need to undertake a different strategy and technique to achieve food security.

COHERENCE IN DEVELOPMENT POLICY KEY TO FOOD SECURITY

The Green Revolution system has shown that increase in yields does not necessarily translate into food security; a "one size fits all" technological strategy does not guarantee food security or even social equity. But it has also shown that when governments act, they can make a difference. The so-called success of the Green Revolution system was due to heavy government intervention in terms of providing subsidies, building infrastructure and providing guarantee for credits. Yet even at the government level, there is lack of policy coherence and sustainability of implementation to sustain agricultural development. This is true not only in Asia but in many other developing regions.

Environmental and natural resource management

Agriculture is based on natural resources, whether soil, water or seeds. Agricultural policies are, however, often disengaged from the management of natural ecosystems that would sustain water supply, prevent erosion of topsoil and provide genetic diversity for crops. Millions of dollars that have been pumped into the Green Revolution system are wasted when environment-associated disasters (floods, landslides, water shortages) occur due to lack of policy coherence between the agriculture and the environment/natural resource sectors. Dams and irrigation systems built as part of the Green Revolution system are rendered useless when forests are allowed to be cleared, causing soil erosion and damage to the water supply system.

Whatever agricultural revolution Africa wants to undertake, it needs to take into account the carrying capacity of the natural resources and adapt to it. As reported by Pingali and Rosegrant (1994), emerging sustainability problems in intensive rice agriculture show the need for a greater understanding of the physical, biological and ecological consequences of agricultural intensification and greater research attention to long-term management of the agricultural resource base.

Industrial and other development policies

In Indonesia, one of the important factors in the decline of rice production over the last few years was reduction in farmlands. Many fertile lands, especially in the more developed islands, have been converted to industrial complexes, tourism facilities or housing facilities. This phenomenon came after the success of the Green Revolution when the government decided to embark on industrial development.

Instead of planning industrial development in areas that are less fertile, it was done in the same areas where the Green Revolution system was developed, mainly on the islands of Java, Sumatra and Bali. As a result, agriculture had to compete with industries for water and land; inevitably agriculture is always the loser. No agriculture revolution can be a success without protection of farmlands.

With industrial development, which is usually more rapid than agricultural development, an imbalance is created between urban and rural development. This leads to urbanization, especially among the younger generation. The Green Revolution could not solve this in Asia or elsewhere, whatever powerful technology is used, since policy coherence is the key.

Social issues

The Green Revolution system is often thought of only as technological innovation. It has, however, a social construct that is against small-scale production and has impacts on food security. For instance, increased production happens more in larger farms because small farmers cannot afford to buy the expensive inputs. In fact, in many countries studies have shown that the Green Revolution has displaced large numbers of smallholder farmers. This has led to increased concentration of land ownership and more intensive urbanization.

The disparity in the distribution of benefits is very clear, even at the national level. The ADB report in 1976 said that the Green Revolution technology covered much of the irrigated areas of Asia, but cannot be easily extended or intensified due to deficiencies in infrastructure and to institutional obstacles (Morgan, 1978). In the case of Indonesia, it was developed well on some of the most developed islands such as Java, Bali, Sumatra and Sulawesi, but not in the eastern part of the archipelago where infrastructure was at a minimum in the 1970s. Thus, the Green Revolution actually increased inequality at the national and at the community levels.

The Green Revolution has also taken away the independence of farmers over management of resources and makes them dependent on external inputs. It has also negated their role as "field scientists", as seed developers and as water managers. Farmers have also become dependent upon government price support and the market system. When this support was dismantled during the economic crisis of East Asia in 1997, farmers found they had to learn to struggle on their own. When farmers wanted to shift to organic integrated agriculture, they found that they had lost many adaptive seeds and the associated knowledge.

Finally, the Green Revolution, because of the mechanical, monoculture, market and yield-based approach, has deprived many communities of their cultural ties to agriculture. Many aspects of local cultures ranging from seed-saving systems to cuisine and even arts were not compatible with the Green Revolution system and had to be abandoned in many areas. This has led to a crisis of cultural identity among many communities.

DIVERSE ALTERATIVES EXIST

Governments in Asia applied the Green Revolution as the only technology for food production during the 1970s, to the exclusion of everything else. Funding (even for research), policy and institutional support were mostly directed towards the Green Revolution. Potential existing technologies and systems, at the farm and at the academic levels, were ignored, or even considered non-existent. These technologies and systems, such as the System of Rice Intensification (SRI), crop rotations, alternative green manures, companion planting and multi-cropping, have now been proven to be viable and sustainable.

Africa can learn from the fact that diversity, at the farm level, cultural level, and technological level, even at the market level, is the key to food security. Diverse agro-ecosystems need diverse approaches. "One size fits all" and "business as usual" scenarios are no longer viable to attain food security and rural development.

CONCLUSION

The lessons of the Green Revolution in Asia can be used by Africa as inputs for considering strategies and approaches to food security. The entire range of lessons, together with knowledge on existing potential technologies and systems in Africa, need to be analysed and considered before any agricultural revolution is undertaken in Africa.

The most vital consideration may be about local agro-ecosystems and what they can offer, rather than applying technologies that are developed detached from the local systems. Agro-ecosystem development may be more important than any revolution. Africa still has that legacy, which can be further improved with appropriate and people-based technology.

The world has changed compared to the 1960s when the Green Revolution was adopted in Asia. We now face multiple crises of natural resource erosion, climate change, globalization and, most recently, the financial crisis. A different approach is needed to develop agriculture and food security for communities and nations. Africa may hold the answer, by learning from the experiences of the Green Revolution in Asia, but also by building on its own strengths of local knowledge, biodiversity and community systems for agriculture and food security.

REFERENCES

Fitzgerald-Moore, P. & B.J. Parai 1996. The Green Revolution. Unpublished. Available at http://docs.google.com/viewer?a=v&q=cache:4uMc21GENfkJ:www.ucalgary.ca/~pfitzger/green.pdf+Fitzgerald-Moore,+P.+and+Parai,+B.J&hl=en&gl=uk&pid=bl&srcid=ADGEESgJxOUsNC_SQu5d7v59_5eYxN7j_7ffPaJn8rQjnpbKnJYKqARk09cjDI7Nx9OZg1H7jRbNJDe-JORFPaZ3_3eNPsiPT-rxryh (accessed 21 August 2009; verified 23 February 2010).

Jeyaratnam, J. 1993. Acute pesticide poisoning in Asia: The problem and its prevention. *In* G. Forget, T. Goodman and A. de Villiers (eds.) *Impact of pesticide use on health in developing countries.* Proceedings of a symposium held in Ottawa, Canada. 17-20 September 1990. International Development Research Centre, Ottawa.

Jhamtani, H. 2008. *Putting food first: Towards community-based food security system.* Insist Press Policy Paper Series. INSISTPress, Yogyakarta.

Lim L.C. 2009. *The case for sustainable agriculture: Meeting productivity and climate challenges.* Environment and Development Series No. 9. Third World Network, Penang.

Morgan, J.P. 1978. The Green Revolution in Asia: False promise of abundance. *Bulletin of Concerned Asian Scholars* 10(1): 2.

Pingali, P.L. & M.W. Rosegrant. 1994. *Confronting the environmental consequences of the Green Revolution in Asia.* Environment and Production Technology Division (EPTD) Discussion Paper No. 2. International Food Policy Research Institute, Washington D.C. Available at http://www.ifpri.org/sites/default/files/publications/eptdp02.pdf (accessed 21 August 2009; verified 23 February 2010).

Sharma, D. 2004. The collapse of Green Revolution. 31 July 2004. Available at http://www.stwr.org/food-security-agriculture/the-collapse-of-green-revolution.html (accessed 21 August 2009; verified 23 February 2010).

©FAO/I. Balderi

INTERNATIONAL ASSESSMENT OF AGRICULTURAL KNOWLEDGE, SCIENCE AND TECHNOLOGY FOR DEVELOPMENT: SUMMARY FOR DECISION MAKERS OF THE SUB-SAHARAN AFRICA REPORT

Carol Markwei,
Lindela Ndlovu,
Elizabeth Robinson and
Wahida Patwa Shah

CONTENTS

ANNEXES

FOREWORD

The objective of the International Assessment of Agricultural Knowledge, Science and Technology for Development (IAASTD) was to assess the impacts of past, present and future agricultural knowledge, science and technology on the:

- o reduction of hunger and poverty,
- o improvement of rural livelihoods and human health, and
- o equitable, socially, environmentally and economically sustainable development.

The IAASTD was initiated in 2002 by the World Bank and the Food and Agriculture Organization of the United Nations (FAO) as a global consultative process to determine whether an international assessment of agricultural knowledge, science and technology was needed. Mr. Klaus Töpfer, Executive Director of the United Nations Environment Programme (UNEP) opened the first Intergovernmental Plenary (30 August-3 September 2004) in Nairobi, Kenya, during which participants initiated a detailed scoping, preparation, drafting and peer review process.

The outputs from this assessment are a Global and five Sub-Global reports; a Global and five Sub-Global Summaries for Decision Makers; and a cross-cutting Synthesis Report with an Executive Summary. The Summaries for Decision Makers and the Synthesis Report specifically provide options for action to governments, international agencies, academia, research organizations and other decision makers around the world.

The reports draw on the work of hundreds of experts from all regions of the world who have participated in the preparation and peer review process. As has been customary in many such global assessments, success depended first and foremost on the dedication, enthusiasm and cooperation of these experts in many different but related disciplines. It is the synergy of these interrelated disciplines that permitted IAASTD to create a unique, interdisciplinary regional and global process.

We take this opportunity to express our deep gratitude to the authors and reviewers of all of the reports—their dedication and tireless efforts made the process a success.

We thank the Steering Committee for distilling the outputs of the consultative process into recommendations to the Plenary, the IAASTD Bureau for their advisory role during the assessment and the work of those in the extended Secretariat. We would specifically like to thank the cosponsoring organizations of the Global Environment Facility (GEF) and the World Bank for their financial contributions as well as the FAO, UNEP, and the United Nations Educational, Scientific and Cultural Organization (UNESCO) for their continued support of this process through allocation of staff resources.

We acknowledge with gratitude the governments and organizations that contributed to the Multi-Donor Trust Fund (Australia, Canada, the European Commission, France, Ireland, Sweden, Switzerland, and the United Kingdom) and the United States Trust Fund. We also thank the governments who provided support to Bureau members, authors and reviewers in other ways. In addition, Finland provided direct support to the Secretariat. The IAASTD was especially successful in engaging a large number of experts from developing countries and countries with economies in transition in its work; the Trust Funds enabled financial assistance for their travel to the IAASTD meetings.

We would also like to make special mention of the Regional Organizations who hosted the regional coordinators and staff and provided assistance in management and time to ensure success of this enterprise: the African Center for Technology Studies (ACTS) in Kenya, the Inter-American Institute for Cooperation on Agriculture (IICA) in Costa Rica, the International Center for Agricultural Research in the Dry Areas (ICARDA) in Syria, and the WorldFish Center in Malaysia.

The final Intergovernmental Plenary in Johannesburg, South Africa was opened on 7 April 2008 by Achim Steiner, Executive Director of UNEP. This Plenary saw the acceptance of the Reports and the approval of the Summaries for Decision Makers and the Executive Summary of the Synthesis Report by an overwhelming majority of governments.

Co-chairs

Hans H. Herren   Judi Wakhungu

Director

Robert T. Watson

STATEMENT BY GOVERNMENTS

All countries present at the final intergovernmental plenary session held in Johannesburg, South Africa in April 2008 welcome the work of the IAASTD and the uniqueness of this independent multi-stakeholder and multidisciplinary process, and the scale of the challenge of covering a broad range of complex issues. The Governments present recognize that the Global and sub-Global Reports are the conclusions of studies by a wide range of scientific authors, experts and development specialists and while presenting an overall consensus on the importance of agricultural knowledge, science and technology for development also provide a diversity of views on some issues.

All countries see these Reports as a valuable and important contribution to our understanding on agricultural knowledge, science and technology for development recognizing the need to further deepen our understanding of the challenges ahead. This Assessment is a constructive initiative and important contribution that all governments need to take forward to ensure that agricultural knowledge, science and technology fulfils its potential to meet the development and sustainability goals of the reduction of hunger and poverty, the improvement of rural livelihoods and human health, and facilitating equitable, socially, environmentally and economically sustainable development.

In accordance with the above statement, the following governments approve the sub-Saharan Africa Summary for Decision Makers:

Benin, Botswana, Cameroon, Democratic Republic of Congo, Ethiopia, Gambia, Ghana, Kenya, Mozambique, Namibia, Nigeria, Senegal, Swaziland, United Republic of Tanzania, Togo, Uganda, Zambia (17 countries).

BACKGROUND

In August 2002, the World Bank and the Food and Agriculture Organization (FAO) of the United Nations initiated a global consultative process to determine whether an international assessment of agricultural knowledge, science and technology was needed. This was stimulated by discussions at the World Bank with the private sector and nongovernmental organizations (NGOs) on the state of scientific understanding of biotechnology and more specifically transgenics. During 2003, eleven consultations were held, overseen by an international multistakeholder steering committee and involving over 800 participants from all relevant stakeholder groups, e.g., governments, the private sector and civil society. Based on these consultations the steering committee recommended to an Intergovernmental Plenary meeting in Nairobi, Kenya, in September 2004 that an international assessment of the role of agricultural knowledge, science and technology in reducing hunger and poverty, improving rural livelihoods and facilitating environmentally, socially and economically sustainable development was needed. The concept of an International Assessment of Agricultural Science and Technology for Development (IAASTD) was endorsed as a multi-thematic, multi-spatial, multi-temporal intergovernmental process with a multi-stakeholder Bureau cosponsored by the Food and Agricultural Organization of the United Nations (FAO), the Global Environment Facility (GEF), United Nations Development Programme (UNDP), United Nations Environment Programme (UNEP), United Nations Educational, Scientific and Cultural Organization (UNESCO), the World Bank and World Health Organization (WHO).

The IAASTD's governance structure is a unique hybrid of the Intergovernmental Panel on Climate Change (IPCC) and the nongovernmental Millennium Ecosystem Assessment (MA). The stakeholder composition of the Bureau was agreed at the Intergovernmental Plenary meeting in Nairobi: it is geographically balanced and multi-stakeholder with 30 government and 30 civil society representatives (NGOs, producer and consumer groups, private sector entities and international organizations) in order to ensure ownership of the process and findings by a range of stakeholders.

About 400 of the world's experts were selected by the Bureau, following nominations by stakeholder groups, to prepare the IAASTD Report (comprised of a Global and five sub-Global assessments). These experts worked in their own capacity and did not represent any particular stakeholder group. Additional individuals, organizations and governments were involved in the peer review process.

The IAASTD development and sustainability goals were endorsed at the first Intergovernmental Plenary and are consistent with a subset of the UN Millennium Development Goals (MDGs): the reduction of hunger and poverty, the improvement of rural livelihoods and human health, and facilitating equitable, socially, environmentally and economically sustainable development. Realizing these goals requires acknowledging the multi-functionality of agriculture: the challenge is to simultaneously meet development and sustainability goals while increasing agricultural production.

Meeting these goals has to be placed in the context of a rapidly changing world of urbanization, growing inequities, human migration, globalization, changing dietary preferences, climate change, environmental degradation, a trend toward biofuels and an increasing population. These conditions are affecting local and global food security and putting pressure on productive capacity and ecosystems. Hence there are unprecedented challenges ahead in providing food within a global trading system where there are other competing uses of agricultural and other natural resources. Agricultural knowledge, science and technology alone cannot solve these problems, which are caused by complex political and social dynamics; but it can make a major contribution to meeting development and sustainability goals. Never before has it been more important for the world to generate and use agricultural knowledge, science and technology.

Given the focus on hunger, poverty and livelihoods, the IAASTD pays special attention to the current situation, issues and potential opportunities to redirect the current agricultural knowledge, science and technology system to improve the situation for poor rural people, especially small-scale farmers, rural labourers and others with limited resources. It addresses issues critical to formulating policy and provides information for decision makers confronting conflicting views on contentious issues such as the environmental consequences of productivity increases, environmental and human health impacts of transgenic crops, the consequences of bioenergy development on the environment and on the long-term availability and price of food, and the implications of climate change on agricultural production. The Bureau agreed that the scope of the assessment needed to go beyond the narrow confines of science and technology (S&T) and should encompass other types of relevant knowledge (e.g., knowledge held by agricultural producers, consumers and end users) and that it should also assess the role of institutions, organizations, governance, markets and trade.

The IAASTD is a multi-disciplinary and multi-stakeholder enterprise requiring the use and integration of information, tools and models from different knowledge paradigms including local and traditional knowledge. The IAASTD does not advocate specific policies or practices; it assesses the major issues facing agricultural knowledge, science and technology and points towards a range of options for action that meet development and sustainability goals. It is policy relevant, but not policy prescriptive. It integrates scientific information on a range of topics that are critically interlinked, but often addressed independently, i.e. agriculture, poverty, hunger, human health, natural resources, environment, development and innovation. It will enable decision makers to bring a richer base of knowledge to bear on policy and management decisions on issues previously viewed in isolation. Knowledge gained from historical analysis (typically the past 50 years) and an analysis of some future development alternatives to 2050 form the basis for assessing options for action on science and technology, capacity development, institutions and policies, and investments.

The IAASTD is conducted according to an open, transparent, representative and legitimate process; is evidence-based; presents options rather than recommendations; assesses different local, regional and global perspectives; presents different views, acknowledging that there can be more than one interpretation of the same evidence based on different world views; and identifies the key scientific uncertainties and areas on which research could be focused to advance development and sustainability goals.

The IAASTD is composed of a Global assessment and five sub-Global assessments: Central and West Asia and North Africa (CWANA); East and South Asia and the Pacific (ESAP); Latin America and the Caribbean (LAC); North America and Europe (NAE); sub-Saharan Africa (SSA). It (i) assesses the generation, access, dissemination and use of public and private sector agricultural knowledge, science and technology in relation to the goals, using local, traditional and formal knowledge; (ii) analyse existing and emerging technologies, practices, policies and institutions and their impact on the goals; (iii) provides information for decision makers in different civil society, private and public organizations on options for improving policies, practices, institutional and organizational arrangements to enable agricultural knowledge, science and technology to meet the goals; (iv) brings together a range of stakeholders (consumers, governments, international agencies and research organizations, NGOs, private sector, producers, the scientific community) involved in the agricultural sector and rural development to share their experiences, views, understanding and vision for the future; and (v) identifies options for future public and private investments in agricultural knowledge, science and technology. In addition, the IAASTD will enhance local and regional capacity to design, implement and utilize similar assessments.

In this assessment agriculture is used in the widest sense to include production of food, feed, fuel, fibre and other products and to include all sectors from production of inputs (e.g., seeds and fertilizer) to consumption of products. However, as in all assessments, some topics were covered less extensively than others (e.g., livestock, forestry, fisheries and agricultural engineering), largely due to the expertise of the selected authors.

The IAASTD draft Report was subjected to two rounds of peer review by governments, organizations and individuals. These drafts were placed on an open access web site and open to comments by anyone. The authors revised the drafts based on numerous peer review comments, with the assistance of review editors who were responsible for ensuring the comments were appropriately taken into account. One of the most difficult issues authors had to address was criticisms that the report was too negative. In a scientific review based on empirical evidence, this is always a difficult comment to handle, as criteria are needed in order to say whether something is negative or positive. Another difficulty was responding to the conflicting views expressed by reviewers. The difference in views was not surprising given the range of stakeholder interests and perspectives. Thus one of the key findings of the IAASTD is that there are diverse and conflicting interpretations of past and current events, which need to be acknowledged and respected.

The Global and sub-Global Summaries for Decision Makers and the Executive Summary of the Synthesis Report were approved at an Intergovernmental Plenary in Johannesburg, South Africa in April 2008. The Synthesis Report integrates the key findings from the Global and sub-Global assessments, and focuses on eight Bureau-approved topics: bioenergy; biotechnology; climate change; human health; natural resource management; traditional knowledge and community based innovation; trade and markets; and women in agriculture.

The IAASTD builds on and adds value to a number of recent assessments and reports that have provided valuable information relevant to the agricultural sector, but have not specifically focused on the future role of agricultural knowledge, science and technology, the institutional dimensions and the multi-functionality of agriculture. These include: FAO State of Food Insecurity in the World (yearly); InterAcademy Council Report: Realizing the Promise and Potential of African Agriculture (2004); UN Millennium Project Task Force on Hunger (2005); Millennium Ecosystem Assessment (2005); CGIAR Science Council Strategy and Priority Setting Exercise (2006); Comprehensive Assessment of Water Management in Agriculture: Guiding Policy Investments in Water, Food, Livelihoods and Environment (2007); Intergovernmental Panel on Climate Change Reports (2001 and 2007); UNEP Fourth Global Environmental Outlook (2007); World Bank World Development Report: Agriculture for Development (2007); IFPRI Global Hunger Indices (yearly); and World Bank Internal Report of Investments in sub-Saharan Africa (2007).

Financial support was provided to the IAASTD by the cosponsoring agencies, the governments of Australia, Canada, Finland, France, Ireland, Sweden, Switzerland, US and UK, the European Commission, and CropLife International. In addition, many organizations have provided in-kind support. The authors and review editors have given freely of their time, largely without compensation.

The Global and sub-Global Summaries for Decision Makers and the Synthesis Report are written for a range of stakeholders, i.e., government policy makers, private sector, NGOs, producer and consumer groups, international organizations and the scientific community.

There are no recommendations, only options for action. The options for action are not prioritized because different options are actionable by different stakeholders, each of whom have a different set of priorities and responsibilities and operate in different socio-economic-political circumstances.

IAASTD SUB-SAHARAN AFRICA REPORT: SUMMARY FOR DECISION MAKERS

Agriculture, which incorporates crops, forests, fisheries, livestock and agroforestry, accounts for an average of 32 percent of the region's GDP, and is woven into the fabric of most societies and cultures in the region. Even though the population is growing and rapidly urbanizing, most families will continue to have ties to land and water.

Agricultural knowledge, science and technology (AKST) has had some notable successes in sub-Saharan Africa including the widespread adoption of improved crop and tree varieties and livestock breeds; the development of pest-resistant and drought-tolerant varieties; biocontrol of pests and parasites such as cassava mealybug, green mite and ticks; integrated natural resource management; development of biodiversity products; and methods and tools for improved productivity and management in water availability for crops, livestock, fodder, trees and fisheries. Yet in sub-Saharan Africa, unlike in other regions, overall per capita agricultural yields declined from 1970 to 1980 and since then have stagnated. The number of poor people is increasing, 30 percent of the population lives with chronic hunger, and similar levels of malnutrition in children under the age of five persist.

Increasing agricultural productivity remains a priority for sub-Saharan Africa, given the very low yields in the region and widespread hunger, poverty and malnutrition. However, the development and sustainable goals of reducing hunger, achieving food security, improving health and nutrition, and increasing environmental and social sustainability will only be reached if the focus of agriculture and agricultural knowledge, science and technology moves away from simply the production of food, fibre, feed, and bioenergy. A broader perspective encompasses an integrated agricultural commodity value chain from production through to processing and marketing with a local and regional perspective. It accounts for the multiple functions of agriculture that include the improvement of livelihoods, the enhancement of environmental services, the conservation of natural resources and biodiversity, and the contribution of agriculture to the maintenance of social and cultural traditions. It recognizes that women, who account for approximately 70 percent of agricultural workers and 80 percent of food processors in sub-Saharan Africa, need significantly increased representation in research, extension and policy making, and equitable access to education, credit and secure land tenure. It also recognizes the need for higher quality education, research and extension that addresses the development and sustainability goals.

Challenges and Options

Current low levels of agricultural productivity in sub-Saharan Africa prevent much of the population from escaping poverty, hunger and malnutrition. On average, livestock and crop yields in sub-Saharan Africa are lower than all other regions, though these averages mask considerable variation. Cereal yields, for example, range from 185 kg ha^{-1} in Botswana to 2 100 kg ha^{-1} in Cameroon. Low yields have been difficult to overcome because they are the result of a wide range of agronomic, environmental, institutional, social and economic factors.

Low input use, including total fertilizer input of less than 10 kg ha^{-1} on average, contributes to sub-Saharan Africa's low crop yields. Although there is considerable variation across farming systems and countries, in the mid-1990s every country in sub-Saharan Africa was estimated to have a negative soil nutrient balance for nitrogen, potassium and phosphorus. Increased fertilizer use is seen by most practitioners as essential, reflected in the resolution by African Union members to reduce costs through national and regional level procurement, harmonization of taxes and regulations, the elimination of taxes and tariffs, and improving access to fertilizer, output market incentives, and credit from input suppliers. The cost of fertilization can also be reduced directly through fertilizer subsidies. These are currently being implemented in some sub-Saharan Africa countries to support farmers. The cost of fertilization can also be reduced through the intensified use of organic fertilizer.

Agrochemicals, especially some synthetic fertilizers and pesticides, have caused negative effects on human and animal health and the environment in some parts of sub-Saharan Africa; this has been exacerbated by unsafe application processes and inadequate access to information concerning handling and disposal practices. Pollution, particularly with respect to water bodies, may also result from inappropriate use. The economic, environmental and health costs associated with greater use of agrochemicals suggest that agricultural knowledge, science and technology options involve reorienting research away from high-input blanket doses towards technologies that enable technically efficient applications specific to local soil conditions and towards integrated nutrient management approaches.

More than four-fifths of agricultural land is affected by soil moisture stress that limits the uptake of nutrients, implying the need to conserve both water and soil organic matter in parallel. Current efforts to improve soil fertility and regenerate the land include research into integrated soil fertility management that builds on farmer practices such as improved natural fallows, rotations, mixed livestock-cropping systems and incorporation of green and livestock manures where available. The adoption of animal manure is limited by transport costs, the quantity needed per unit area of land and labour costs of weeding. Green manures help to revive degraded land, but often compete with food and cash crops, and the benefits are often unnoticed in the short run. These are the types of tradeoffs that agricultural knowledge, science and technology needs to evaluate and minimize with farmers. Organic, agroforestry and no- or low-till farming offer integrated agro-ecological approaches to reducing soil degradation, but further studies are required to determine the conditions and incentives required for farmers to adopt these methods.

Increases in the exploitation of both surface and groundwater are required for sub-Saharan Africa to increase productivity. Agricultural production in sub-Saharan Africa is predominantly rainfed. Only 4 percent of agricultural land is irrigated compared to 37 percent in Asia and 15 percent in Latin America. This situation is exacerbated by high rainfall variability and uncertainty, especially in arid and semiarid areas, and projected rising temperatures in sub-Saharan Africa and decreased precipitation in the Sahel and southern Africa as a consequence of climate change. The characteristics of agriculture in sub-Saharan Africa suggest that smaller-scale irrigation, 'greenwater' technologies such as water conservation, rainwater harvesting and community level water management need to be explored as alternatives to large-scale irrigation projects. Increases in the level of irrigation can come from both surface and ground water, drawing lessons from within and outside the region on viable small to medium scale irrigation techniques that require limited infrastructural development and can reach many farmers. Methods such as pumping from the rivers on an individual and small group basis, and locally manufactured drip systems are still to be fully exploited.

Efficient and equitable water allocation, a component of agricultural knowledge, science and technology, requires a better understanding of the value of water for different competing users, appropriate mechanisms for allocating water, (e.g. pricing, allocation of property rights, regulation) and negotiations that create incentives for farmers to adopt water-efficient technologies. The appropriate approach will require integrated research that builds on local knowledge, existing technologies, existing water institutions and the ability to enforce rights through formal systems, and also on complementary institutions such as land rights and farmers' access to credit. Poor households may simply not be able to afford water priced at its true cost, in which case approaches such as that taken in South Africa (households get a free allocation per month) need to be explored.

Increasing the performance of agriculture requires an improvement in productivity on the 80 percent of sub-Saharan Africa farms that are smaller than two hectares. Earlier paradigms that typically attempted to fit farmers into the existing linear top-down structures of research-development-extension worked relatively well for major cash crops, but there has been less success on small-scale diversified farms. Options for agricultural knowledge, science and technology include integrated and participatory approaches that can increase the likelihood that appropriate technologies for production are developed and adopted by small-scale farmers. Alternative approaches include moving farmer engagement closer to priority setting and funding decisions, increasing collaboration with social scientists, and increasing participatory and interdisciplinary work in the core research institutions. There is evidence from East Africa that innovative approaches to agricultural knowledge, science and technology development such as farmer research groups are more successful in reaching women farmers than traditional extension activities. By understanding farmers' contexts and priorities, grounding new technologies in an understanding of farmers' motivations and constraints, and explicitly including groups that are often socially excluded such as women and minorities, agricultural knowledge, science and technology is more likely to be relevant and adopted.

Many farmers in sub-Saharan Africa use indigenous animal breeds which are able to withstand harsh conditions and tolerate many diseases, but their meat, milk and egg productivity is low. Options for agricultural knowledge, science and technology to improve livestock productivity include the use of open nucleus breeding schemes and improving the genetic potential of indigenous breeds, e.g. through characterizing genetic diversity in order to provide insights into genetic relationships. Given that animal disease management is one of the key explanations for movements, herd size and growth, agricultural knowledge, science and technology has a role to play in addressing the impact of disease at the smallholder level.

Scaling up integrated approaches is difficult because successful innovations tend to incorporate local knowledge and to be specific to the particular agro-climatic conditions. Public good aspects of baskets of prototype technologies, whether originating from farmers, researchers or collaborative efforts, that match the diversity of farmers' fields can be transferred with appropriate scaling up and dissemination strategies. Where current structures are ineffective, new institutional and organizational arrangements may be required to support the empowerment of local communities to develop, adapt and disseminate agricultural knowledge, science and technology. Despite the increasing use of participatory and integrated approaches to agricultural knowledge, science and technology development, institutional resources still tend to be compartmentalized. For example, water management is often undertaken independently of pest, soil, livestock and forest management. Reduced water availability is the main cause of loss of productivity in more than half of the grazing land. Improved water management would improve livestock health through quantity and quality of grazing resources and reduced walking distance to watering points.

Knowledge, understanding and uptake of new agricultural technologies on the whole are poor and patchy in sub-Saharan Africa. In the IAASTD assessment, biotechnology is defined according to that in the Convention on Biological Diversity. In this context it includes much of the traditional knowledge and many of the traditional technologies used in sub-Saharan Africa for the production, processing and preservation of food plus modern molecular tools such as genetic engineering, marker assisted selections or breeding and genomic techniques. In this broader sense biotechnology, as an agricultural knowledge, science and technology subset, has a role to play in addressing development and sustainability goals but it needs to be managed to avoid derivative problems from its use.

Genetic engineering is considered by some to have important ramifications for productivity but some of its uses and impacts are hotly contested. Contamination of farmer-saved seed and threats to biodiversity in centers of origin are key concerns with respect to biotechnology and genetic engineering in particular. The environmental risks and evidence of negative health impacts mean that sub-Saharan Africa's ability to make informed decisions regarding biotechnology research, development, delivery and application is critical. In part, the current limited capacity of individual countries to address risk assessment and management of transgenics is being addressed through regional capacity building and harmonization of guidelines, policies, legislation and creating an understanding of biosafety issues. However, individual countries could develop and advance their own biotechnology capacities. The development of comprehensive national biosafety frameworks must work in conjunction with effective enforcement institutions and implementation mechanisms.

Biological control is an option for integrated pest management and involves augmentation or conservation of local or introduced natural enemies to pest populations. There are several examples where staple and important crops have been saved by biological control over wide areas. There are a number of economic assessments showing biocontrol's successes including coffee mealybug and more recently the campaigns against cassava mealybug, green mite and water hyacinth that show large and accruing gains. These controls are still in place and contribute to small farmers' food security in the long term.

Sub-Saharan Africa countries are the most intense users of biomass in the world, meeting more than 50 percent of their total primary energy consumption from this source. This biomass energy predominantly consists of unrefined traditional fuel such as firewood and crop and animal residues. Use of biomass as a source of energy in its traditional forms results in inefficient energy conversion, environmental and health hazards, is time-consuming in terms of collection and contributes to the degradation of forests. Agricultural knowledge, science and technology has played a role in improving the traditional bioenergy technologies, such as design and supply of fuel-efficient cooking stoves, and helping people to move to more sustainable, efficient and less harmful forms of energy. Some sub-Saharan Africa countries have realized this potential and have programs for the cogeneration of electricity.

R&D in improving biofuel yields per unit of land and in reducing economic costs of production are needed. Biofuel production involves tradeoffs that have not yet been evaluated. Globally, output from first generation biofuels produced from agricultural crops is growing rapidly supported by government policies, but these fuels are rarely economically competitive with petroleum fuels. The production of first generation biofuels in particular in sub-Saharan Africa is likely to put pressure on forests and marginal lands. A major debate centres around whether this use of biomass will remove land from production of food crops and/or result in increased prices of staple commodities, such as maize, if used for biofuels. Next generation biofuels may have greater potential for sub-Saharan Africa. Many use residues, stems and leaves and so could reduce pressure on land requirements, but concerns remain, e.g. over the environmental impact of harvesting agricultural residues. Agricultural knowledge, science and technology has a large role to play concerning the careful analysis of biofuel technology appropriate for sub-Saharan Africa, in parallel with the development of policies and capacity building to reduce the negative effects of growing biofuels and determine the health, environmental, energy and food security tradeoffs in the region. Increased research will also enable sub-Saharan Africa countries to determine their appropriate entry points.

Rapid depletion of sub-Saharan Africa's natural resources and the genetic erosion of indigenous germplasm threaten the sustainability of agriculture in sub-Saharan Africa. Land use change, including deforestation and expansion of agriculture into marginal areas, results in nutrient and biodiversity losses, water and soil degradation, loss of pasture, adversely affects ground and surface water availability and reduces the resilience of agricultural systems, especially in semiarid regions.

These issues affect every aspect of agricultural knowledge, science and technology as environmental degradation affects the productivity and sustainability of agriculture. Over-exploitation of freshwater and oceanic fisheries, controlled breeding and the development of livestock, crop and tree breeds with a narrow genetic base further threaten the resource base.

Integrated natural resource management options include diversifying farming systems, enhancing natural capital and building on local and traditional knowledge. For instance, significant investments have been made in the development of high value products from indigenous plant species for the pharmaceutical, neutraceutical and cosmetic industries. Such localization approaches place agriculture squarely in the context of society and ecosystems and so can empower local communities to address depletion of natural resources and loss of biodiversity, in conjunction with combating poverty and improving food security. Integrated approaches allow the generation of substantive knowledge concerning the trade-offs among economic, social, cultural and ecological goals, the roles of various actors such as producers, the private sector, civil society and government, and can accommodate new challenges such as changes caused by climate change, including the increased problem of invasive species. These sets of activities and interventions will not reach system level goals without an explicit analysis of who wins and who loses and how the potential tradeoffs and synergies will be managed. Strategies of rapid agricultural development need to be coordinated more directly with strategies for biodiversity and water conservation such as retaining areas of natural vegetation in production areas, keeping areas where pollinators can thrive, promoting organic agriculture and incorporating trees in agricultural landscapes.

The public good nature of many natural resources lends itself to consultative and collective approaches in the development of policies and institutions. Involving local communities in determining land use and land tenure policies and giving them control and responsibility over the resources increases the likelihood of efficient, equitable and sustainable use of common pool natural resources and compliance with rules and regulations. Examples include participatory forest management, which is being introduced in a number of countries in sub-Saharan Africa. The collective, public goods aspect of on-farm agricultural biodiversity can be supported through international mechanisms such as Farmers' Rights provisions under the FAO International Treaty on Plant Genetic Resources for Food and Agriculture.

Farmers in sub-Saharan Africa often integrate trees on their farms and on landscapes in order to harness multiple benefits, including timber and other high value products, fuel wood, fibre, feed, medicinal products, fruits and ecosystem services, such as land rehabilitation and soil fertility through sequential fallow systems and systems with intercropped trees. Barriers to clonal forestry and agroforestry have been overcome by the development of robust vegetative propagation techniques, which are applicable to a wide range of tree species. Domestication, intensive selection and conventional breeding have had positive impacts on yield and the production of staple food crops, horticultural crops and timber trees. Agroforestry research builds on local knowledge and has the potential to reduce pressure on forests and provide ecosystem services such as biodiversity conservation, carbon sequestration and land restoration. Women and men have different priorities, which suggest scope for agricultural knowledge, science and technology to identify trees with multiple uses. Factors that need to be taken into account in agroforestry research include impact assessments, e.g., ensuring that trees do not jeopardize water supplies, especially in dry areas, and that exotic species are not introduced that cause social equity issues relating to land use and land rights. Other issues that need to be addressed include increasing adoption of agroforestry technologies, pests and diseases, markets for agroforestry products, availability of planting materials and adaptation to climate change.

Because livestock genetic diversity is being lost relatively rapidly, short-term strategies are required to provide information for priority setting. This might include as a first step, rapid surveys and population estimates and data on genetic distances. In the longer term, policies and market strategies to promote the use of indigenous breeds can provide economic incentives to conserve these breeds.

Community participation in livestock breeding increases the likelihood of appropriate traits being identified and developed. Yet information is still required with respect to how livestock owners make livestock selections and how livestock production fits with other livelihood activities.

Sub-Saharan Africa is the only region where per capita fish supplies are falling (from 9 kg per person in 1973 to 6.6 kg in 2005) as a result of stagnation in capture fish production and a growing population. Where capture fisheries are over-exploited, institutions need to be strengthened for allocating fishing rights, ensuring sustainable catches, and enforcing rules and regulations. Improved management of capture fisheries will also require strategies to reduce and use by-catch, and reduce postharvest losses. Working with local fishing communities and understanding their perspectives on externally enforced rules and regulations may reduce tensions between biological realities and community acceptance. Investment in supporting local fishers in modern fishing techniques could also go a long way in reducing tensions and improving livelihoods.

Unlike in other regions, aquaculture currently makes a very small contribution to total fish production in sub-Saharan Africa – just 2 percent compared with 38 percent worldwide. Aquaculture has the potential to improve livelihoods and nutrition, and reduce the pressure on capture fisheries. Agricultural knowledge, science and technology has a role to play in reducing the potential negative effects of aquaculture through learning from other regions, increased research into integrated farming systems that avoid using wild-caught fish as feed, and strengthening the capacity for impact monitoring, such as the impacts of chemical inputs and the conversion of mangroves to fisheries. Additional options for agricultural knowledge, science and technology include the need to develop post-harvest technologies, value chain and product development, farmer training and increasing access to inputs.

Agricultural intensification tends to be accompanied by decreasing agricultural biodiversity. However, farmers naturally play a role in conserving agricultural biodiversity that can be exploited and incorporated into more formal conservation approaches. Genetic erosion is of particular concern in sub-Saharan Africa because many countries have a wide range of crops and livestock species that are considered relatively unimportant on a global level but are important as local staples. *In situ* conservation and protection is particularly important for conserving genetic resources, helping to maintain evolutionary processes and having a positive effect on biodiversity and equity.

Working with local communities has been shown to be key to conserving biodiversity and maintaining or enhancing ecosystem services in the long term. Market-oriented incentives enable local communities to benefit financially from sustainably managing soils, water, sequestering carbon and conserving biodiversity. These could include direct payments to farmers or to particular agricultural sectors; other types of rewards include well-defined property rights over natural resources in favour of local communities; the development of markets for indigenous species; and strengthening intellectual property rights.

Agriculture, health and nutrition in sub-Saharan Africa are closely linked. The emphasis of agricultural policies in sub-Saharan Africa on the production of a few staple food crops to the neglect of indigenous species with good nutritional properties, and micronutrient rich foods, such as fruits and vegetables, has reduced agriculture's potential to improve the livelihoods of households, including health and nutrition.

Increasing yields will have a direct impact on the nutritional status of the rural poor. General options to reduce malnutrition encompass increasing households' access to income and calories as well as encouraging a diet of diversified foods with the needed nutrients. There is scope for agricultural knowledge, science and technology to target micronutrient deficiency through increased research into the nutritional value of local and traditional foods, particularly fruits and vegetables, and the extent to which they contribute to diets. To ensure that the direction of agricultural knowledge, science and technology research is relevant to local communities and that its outputs will be widely adopted, additional research is required into the conditions under which farmers will choose to cultivate and market these traditional food sources and households will choose to consume and purchase. The empowerment and increased involvement of women can help with the development, adoption and demand for more nutritious foods, such as orange-flesh sweet potato (*Ipomoea batatas*). Malnutrition is increasingly becoming an urban as well as rural problem. Options that are particularly relevant to the urban population include product development to increase the variety and quality of food, including fortified foods, and targeted information campaigns to increase awareness and encourage adoption of more nutritious foods.

Malnutrition and ill health in sub-Saharan Africa are exacerbated by tropical diseases, such as malaria and schistosomiasis, and by HIV/AIDS-associated diseases, such as tuberculosis, that result in a reduced workforce available to agriculture and other productive sectors. Animal-linked diseases affecting both human and animals have also been a significant setback to livelihood security, aggravated by unregulated cross-border movements resulting in the spread of transboundary diseases such as Contagious Bovine Pleuropneumonia (CBPP), African Swine Fever (ASF) and Rift Valley Fever (RVF). Agricultural knowledge, science and technology options to address these diseases include efficient vaccine development, rapid and accurate diagnostic techniques and breeding of animals with high tolerance to diseases. Policy options include control of animal movements across boundaries and this requires regional cooperation.

Most farmers in sub-Saharan Africa operate in an environment of high risk and uncertainty. Farmers therefore tend to adopt strategies that minimize risk and vulnerability at the expense of profit-maximizing strategies, resulting in an agricultural sector in sub-Saharan Africa that is well below its potential. Sub-Saharan Africa already experiences high variability in rainfall and other climatic extremes, which will be exacerbated by climate change. Resilience in much of sub-Saharan Africa is inhibited by fragile ecosystems, weak institutions, ineffective governance, and poverty; those most vulnerable are the poor who have the least adaptive capacity. When agricultural knowledge, science and technology builds on farmers' and pastoralists' coping strategies and innovations thereby placing local people's knowledge and actions, such as diversified production practices used by 90 percent of sub-Saharan Africa farmers, at the centre of research efforts, the multiple functions of agriculture are better realized and the threats of climate change mitigated. Options include undertaking collaborative research with farmers, including the integration of crop, livestock, tree and fish components where applicable that spread risk and deliver various benefits at different periods throughout the year.

Few households in sub-Saharan Africa have private and transferable property rights to the land that they farm. Although secure land tenure correlates with long-term investments in natural resource management, land titling in itself has not been shown to increase credit transactions, improve production or increase the number of land sales. Any benefits are often offset by the high transactions costs of titling land and loss of rights of disadvantaged groups including women and pastoralists. However, land tenure reform in some cases may be necessary to secure individual or collective rights to resources in order to reduce farmers' vulnerability and strengthen women's access to resources. It is more likely to be effective and equitable if it is sensitive to the impact on the rights of disadvantaged groups and undertaken in parallel with the harmonization of other laws such as those governing inheritance. Collective action when resource and land tenure are secure has yielded benefits and reduced risks and costs for members through labour efficiencies, provision of public services and management of natural resources. The inclusion of a gender perspective in these institutions for collective action leads to more equitable outcomes.

Credit, insurance, and other risk-sharing institutions can reduce farmer exposure to risk and uncertainty and therefore enable them to increase expected output and profits. Microcredit is relatively well established in sub-Saharan Africa. Much is provided through NGOs and not all may be economically sustainable without the injection of external funds to cover the relatively high administrative costs. Recently retail banks are becoming involved in commercially viable microcredit by providing capital to organizations that then provide the microcredit directly to farmers. An appropriate policy environment for easy access to affordable microcredit is most likely to benefit farmers. Alternatives to credit from the financial sector include the development of contracts that allow for advanced payment and provision of inputs and extension services from agribusiness companies to farmers, such as contract farming and outgrower schemes.

Weather insurance can reduce farmers' exposure to highly variable rainfall and hence crop yields provided they are in a position to pay for such services. Private provision of weather and crop insurance is only likely to occur for larger farms and high value crops. Some initiatives are being piloted by the World Bank that pay out depending on rainfall rather than crop output, thereby eliminating moral hazard (farmers may put less effort into their farming activities if they are insured against losses). Such insurance may be more relevant to drought rather than climate variability, but the problem of covariance remains (if one farmer is negatively affected the likelihood is that most farmers in the vicinity will be), suggesting that private companies on their own may not be willing to provide such insurance. Micro-insurance is already being introduced for small-scale farmers in a number of sub-Saharan Africa countries through partnerships between private companies, donor governments, and NGOs, but has not been rigorously evaluated.

Rangeland management approaches practiced by pastoral livestock farmers have been recognized as the appropriate response to knowledge of the spatial and temporal availability of resources. These strategies include movement of livestock to follow quality and quantity of feed and water, flexible stocking rates and herd diversification sustained by a system of communal resource tenure. Agricultural knowledge, science and technology needs to address emerging constraints and new realities for these pastoral systems brought about by land tenure changes, which conflict with traditional tenure, institutions, and carrying capacity in the context of emerging challenges such as climate change and associated stresses. These strategies are most likely to work if countries develop regional strategies to enhance the evolution of pastoral farming systems.

Options for agricultural knowledge, science and technology include the application of geographic information systems and quantitative modelling processes to provide further insights into productivity patterns of the system and offer policy options to ensure sustainability. Incentives and arrangements for local communities that designate rangelands for other uses such as biodiversity conservation have been attempted in some countries. The development of reliable early warning systems to avoid catastrophic effects of droughts and designing livestock management systems can help to alleviate the shortage of dry season grazing. Improving understanding and documentation of the role of livestock in livelihoods and motivations behind pastoralist practices will be most effective if conducted in pastoralists' languages using participatory methods.

The lack of connection between sub-Saharan Africa farmers and the market has seen agriculture remain rudimentary, unprofitable and unresponsive to market demand. Farmers' poor access to markets reduces incentives to apply agricultural knowledge, science and technology innovations and to make investments in modern technologies and so inhibits the shift of poor farmers from subsistence to market-oriented production. Weak markets result in expensive inputs and poorly developed output markets result in low farm-gate prices for internationally traded products. Weak business service sectors reinforce small producers' isolation from any but the most local markets and barriers to entering the formal market reinforce the inefficiencies and limitations inherent in the informal sector, with the result that the benefits of informality are outweighed by reduced competitiveness and increased vulnerability. Sub-Saharan Africa farmers have fared no better internationally. Between 1980 and 2000, most sub-Saharan Africa countries' agricultural exports to international markets stagnated at just 2 percent of the global market in spite of globalization trends that were expected to open new markets to sub-Saharan Africa products. It is critical that terms of trade between sub-Saharan Africa and international partners improve.

Options to improve the connection between farmers and the market include improving technical assistance in production and post-harvest techniques; training and capacity development and access to credit for long-term investments and product upgrading; investment in organizational and institutional development of farmer organizations to enhance farmers' management, negotiating, and bargaining skills; and promotion of agro-processing in small urban centres. Agricultural knowledge, science and technology has an important role to play in increasing production efficiency along the value chain by making modern technologies available and providing viable processes for transmitting marketing information and including information related to consumer preferences and price signals to farmers and agro-processors. Contract farming and outgrower schemes, which offer benefits related to guaranteed market access, access to credit and market information are being explored in the region.

The absence of processing and storage infrastructure located near the main producing areas inhibits value addition. Further, market development calls for infrastructure inputs, including rural road networks and electricity. There is a positive correlation between the development of transportation infrastructure and agricultural intensification; yet sub-Saharan Africa has the lowest density of paved roads of any world region. Information and communication technologies (ICTs) development is increasing access to and contribution of agricultural knowledge, science and technology knowledge in some parts of the region, but there is potential to achieve more impact.

Increasing the scope of marketing opportunities at the regional level, as stipulated in the Lagos Plan of Action and the Abuja Treaty, will increase trade and marketing opportunities. Further options include implementing existing regional agreements towards meeting targets; improving and harmonizing customs procedures and instituting policies for more efficient cross-border trade; and removing infrastructural and other barriers to the movement of commodities across borders.

Payments for environmental services (PES) are a market-based tool that has received substantial interest in sub-Saharan Africa. It creates incentives for managing natural resources, directly rewarding management practices that contribute to maintaining and enhancing environmental services that result in biodiversity conservation, carbon sequestration, water quality and availability, and land rehabilitation and nutrient cycling. There has been some recent experience in sub-Saharan Africa where those that provide an environmental service are compensated for this by the beneficiaries of the service.

There is also increasing potential for African countries and small-scale farmers to be involved in voluntary markets for carbon and international market mechanisms such as the CDM (Clean Development Mechanism). Knowledge and strategies to reduce carbon emissions through community based afforestation and reforestation projects, agroforestry and reduced deforestation and degradation (REDD) are being generated, but need to be tested and adopted/adapted. These strategies have the potential to create synergies for increasing productivity and achieving the multiple functions of agriculture.

Other mechanisms such as certification, which may result in a premium paid to farmers, have to be carefully designed so that appropriate prices are set and the requirements for certified products are jointly negotiated. However, at present the costs of certification for small-scale farmers can be prohibitive. Agricultural knowledge, science and technology has a role to play in assessing and monitoring the impacts of these different, novel market approaches – decreasing transactions costs for local communities, and setting up appropriate policies and institutions that provide level playing fields for negotiation between buyers and sellers and determine whether the poor can benefit.

The dominance of external funding for agricultural knowledge, science and technology in sub-Saharan Africa has resulted in unreliable long-term funding and loss of control over the relevance and direction of new technology developments. Even with external funding, if Nigeria and South Africa are excluded, agricultural knowledge, science and technology spending in sub-Saharan Africa declined by 2.5 percent per year during the 1990s. A commitment by countries in sub-Saharan

Africa to reaching the Maputo Declaration's target of allocating 10 percent of the budget to agriculture has the potential in some cases to ensure more sustained and reliable public funding for agricultural knowledge, science and technology, increase its relevance for sub-Saharan Africa, and be a catalyst for increased coherence between donor and national policies. In parallel, better use can be made of current limited resources through existing regional and sub-regional networks enabling resource and expertise sharing; leveraging funding through cost-sharing with end users; the use of competitive grants, matching grants, trust funds, and specific surcharges such as levies and voluntary contributions. Furthermore, a strategic action at the national level on stimulating local private sector investment in food and agriculture and local agri-business could help.

Establishing funding mechanisms through performance-based competitive research funds and matching grants can enhance collaboration between various research partners. Public-private partnerships (PPP) offer a way to leverage public funding, but agricultural knowledge, science and technology research and development may be pulled towards commercial outputs at the expense of public good outputs and so still need to be evaluated against development and sustainability goals. Given the contribution of agriculture to improving human health and nutrition, a strategy of integrated planning and programming among ministries of health, agriculture, livestock and fisheries would provide opportunities for joint funding of, and better synergies among programs. More generally, shifting to a multifunctional localized approach to agriculture will require political will on the part of policy makers, agribusinesses and donors of publicly funded research to make more community-centred decisions about how to invest limited resources.

Current education, training and extension structures are incompatible with innovative approaches to agricultural knowledge, science and technology development. Most agricultural scientists in sub-Saharan Africa are trained and rewarded within a narrow discipline, reflecting the typically linear approaches to research and extension that value "formal" scientific research and learning over more tacit forms of farmer learning and local and traditional knowledge. Proven approaches to research for development have evolved recently, with more attention paid to integrated solutions, spatial heterogeneity, tradeoffs, and livelihood and environment outcomes rather than only productivity issues. There has also been considerable emphasis in establishing coherence and synergies among basic applied and adaptive research as well as dissemination of results by encouraging collective participation of universities, private sector, public research organizations and civil society. New players, including some international NGOs, have joined in knowledge generation.

In sub-Saharan Africa, the generation of formal knowledge and scientific development rests predominantly with a research system comprising national and international agricultural research organizations, universities and the private sector. Often this research system is slow and inadequate in its response to challenges. This is partly due to poor access to current global literature and expertise. Typically it can also be attributed to education systems that inadequately prepare scientists to carry out effective research, and to poor linkages between education, research and extension. Education is still centred on learning facts rather than developing skills in problem solving and is constrained by disciplinary boundaries.

Options include improving the connections between education, research and extension systems, moving to problem-based learning, removing outdated disciplinary paradigms and updating the research approaches and tools being taught. Training can be expanded to include the socioeconomic and policy environment in which agricultural development occurs, and field-based research with farmers. A new cadre of specialists is needed who are able to offer technical support in appropriate tools and approaches. However, scientists are less likely to choose to undertake longer-term participatory and integrated research unless there are changes in the professional reward system that is currently based predominantly on the generation of data at meso and macro levels.

There is scope to explore the potential for efficiencies in regional graduate training models. The large number of small countries in Africa means it is often difficult for individual universities to achieve a critical mass of teachers in specialized areas such as biotechnology. Appropriately designed regional training approaches may provide a solution. However, rather than creating new regional institutions, self-initiated efforts—building on regional specializations within existing universities and then developing networked training programs to attract students from a regional watershed—are likely to be more cost effective and have more impact, particularly in the short term.

New approaches to agricultural knowledge, science and technology generation that increase farmer involvement and include local and traditional knowledge naturally incorporate and enhance farmers' own technical skills and research capabilities. However, sub-Saharan Africa is the only region where formal education and government services function formally in languages different from the first languages of almost the entire citizenry. This linguistic divide, which reduces the scope for combining formal science and technology and local and traditional knowledge, can be addressed in part through the increased use and understanding of local languages when working with farmers.

Increasing the functional literacy and general education levels among rural communities, especially women, has already been proven to increase the likelihood of achieving development and sustainability goals. Additional options include specific curriculum reform that addresses the key skills required to empower individuals and communities to engage in the development and use of agricultural knowledge, science and technology, increase the likelihood of local and traditional knowledge being incorporated, and drive and contribute to agricultural product and service development. Specific actions to mainstream women's involvement include strategies that encourage women to study agricultural and engineering sciences and social sciences; and effort to ensure that extension, data collection and enumeration involve women both as providers as well as recipients. For example, 83 percent of extension officials in sub-Saharan Africa are men who, due to cultural norms cannot, or may choose not to speak to women.

ANNEX A
AUTHORS AND REVIEW EDITORS

BENIN
Peter Neuenschwander
*International Institute of
Tropical Agriculture*
Simplice Davo Vodouhe
Pesticide Action Network

DEMOCRATIC REPUBLIC OF CONGO
Dieudonne Athanase Musibono
University of Kinshasa

ETHIOPIA
Assefa Admassie
*Ethiopian Economic Policy
Research Institute*
Gezahegn Ayele
*Ethiopian Development Research
Institute-International Food policy
Research Institute (EDRI-IFRPI)*
Joan Kagwanja
Economic Commission for Africa
Yalemtsehay Mekonnen
Addis Ababa University

GHANA
John-Eudes Andivi Bakang
*Kwame Nkrumah University of
Science and Technology (KNUST)*
Daniel N. Dalohoun
*United Nations University
MERIT- Institut National de recherche
agronomique (MERIT-INRA)*
Felix Yao Mensa Fiadjoe
University of Ghana
Carol Mercey Markwei
University of Ghana
Legon Joseph (Joe) Taabazuing
*Ghana Institute of Management and
Public Administration (GIMPA)*

KENYA
Tsedeke Abate
ICRISAT-Nairobi

Susan Kaaria
Ford Foundation
Washington O. Ochola
Egerton University
Wellington Otieno
Maseno University
Wahida Patwa Shah
World Agroforestry Centre (ICRAF)
Anna Stabrawa
*United Nations Environment
Programme (UNEP)*

MADASCAGAR
Roland Xavier Rakotonjanahary
*National Center for Applied
Research for Rural Development
(FOFIFA)*

MOZAMBIQUE
Manuel Amane
*Instituto de Investigação
Agrícola de Moçambique
(IIAM)*
Patrick Matakala
*World Agroforestry Centre (ICRAF)
Southern Africa Regional Programme*

NETHERLANDS
Nienke Beintema
*International Food Policy
Research Institute (IFPRI)*

NIGERIA
Sanni Adunni
Ahmadu Bello University
Michael Chidozie Dike
Ahmadu Bello University
V.I.O. Ndirika
Ahmadu Bello University
Stella Williams
Obafemi Awolowo University

RWANDA
Agnes Abera Kalibata
Ministry of Agriculture

SENEGAL
Julienne Kuiseu
*West and Central African Council for
Agricultural and Research Development
(CORAF/WECARD)*
Moctar Toure
Independent

SOUTH AFRICA
Marnus Gouse
University of Pretoria

SRI LANKA
Francis Ndegwa Gichuki
*International Water Management
Institute (IWMI)*

TANZANIA
Roshan Abdallah
*Tropical Pesticides Research Institute
(TPRI)*
Stella N. Bitende
*Ministry of Livestock and Fisheries
Development*
Sachin Das
*Animal Diseases Research Institute
(ADRI)*
Evelyne Lazaro
Sokoine University of Agriculture
Razack Lokina
University of Dar es Salaam
Lutgard Kokulinda Kagaruki
*Animal Diseases Research Institute
(ADRI)*
Elizabeth J.Z. Robinson
University of Dar es Salaam

UGANDA
Apili E.C. Ejupu
*Ministry of Agriculture, Animal
Industries and Fisheries*
Apophia Atukunda
Environment Consultancy League (ECL)

Dan Nkoowa Kisauzi
*Nkoola Institutional Development
Associates (NIDA)*
Imelda Kashaija
*National Agriculture Resource
Organization (NARO)*

UNITED KINGDOM
Nicola Spence
Central Science Laboratory

UNITED STATES
Wisdom Akpalu
*Environmental Economics
Research & Consultancy
(EERAC)*
Patrick Avato
The World Bank (WB)
Mohamed Bakarr
*Center for Applied Biodiversity
Science, Conservation International*
Amadou Makhtar Diop
Rodale Institute
David Knopp
Emerging Markets Group (EMG)
Pedro Marques
The World Bank (WB)
Harry Palmier
The World Bank (WB)
Stacey Young
*US Agency for International
Development (USAID)*

ZAMBIA
Charlotte Wonani
University of Zambia

ZIMBABWE
Chiedza L. Muchopa
University of Zimbabwe
Lindela R. Ndlovu
*National University of Science and
Technology*
Idah Sithole-Niang
University of Zimbabwe

ANNEX B
SECRETARIAT AND COSPONSOR FOCAL POINTS

SECRETARIAT

World Bank
Marianne Cabraal
Leonila Castillo
Jodi Horton
Betsi Isay
Pekka Jamsen
Pedro Marques
Beverly McIntyre
Wubi Mekonnen
June Remy
UNEP
Marcus Lee
Nalini Sharma
Anna Stabrawa
UNESCO
Guillen Calvo

COSPONSOR FOCAL POINTS

GEF	Mark Zimsky
UNDP	Philip Dobie
UNEP	Ivar Baste
UNESCO	Salvatore Arico, Walter Erdelen
WHO	Jorgen Schlundt
World Bank	Mark Cackler, Kevin Cleaver, Eija Pehu, Juergen Voegele

REGIONAL INSTITUTES

Sub-Saharan Africa – African Centre for Technology Studies (ACTS)
Ronald Ajengo
Elvin Nyukuri
Judi Wakhungu
Central and West Asia and North Africa – International Center for Agricultural Research in the Dry Areas (ICARDA)
Mustapha Guellouz
Lamis Makhoul
Caroline Msrieh-Seropian
Ahmed Sidahmed
Cathy Farnworth
Latin America and the Caribbean – Inter-American Institute for Cooperation on Agriculture (IICA)
Enrique Alarcon
Jorge Ardila Vásquez
Viviana Chacon
Johana Rodríguez
Gustavo Sain
East and South Asia and the Pacific – WorldFish Center
Karen Khoo
Siew Hua Koh
Li Ping Ng
Jamie Oliver
Prem Chandran Venugopalan

ANNEX C
STEERING COMMITTEE FOR CONSULTATIVE PROCESS AND ADVISORY BUREAU FOR ASSESSMENT

STEERING COMMITTEE

The Steering Committee was established to oversee the consultative process and recommend whether an international assessment was needed, and if so, what was the goal, the scope, the expected outputs and outcomes, governance and management structure, location of the Secretariat and funding strategy.

CO-CHAIRS

Louise Fresco
Assistant Director General for Agriculture, FAO
Seyfu Ketema
Executive Secretary, Association for Strengthening Agricultural Research in East and Central Africa (ASARECA)
Claudia Martinez Zuleta
Former Deputy Minister of the Environment, Colombia
Rita Sharma
Principal Secretary and Rural Infrastructure Commissioner, Government of Uttar Pradesh, India
Robert T. Watson
Chief Scientist, The World Bank

NON GOVERNMENTAL ORGANIZATIONS

Benny Haerlin
Advisor, Greenpeace International
Marcia Ishii-Eiteman
Senior Scientist, Pesticide Action Network North America Regional Center (PANNA)
Monica Kapiriri
Regional Program Officer for NGO Enhancement and Rural Development Aga Khan

Raymond C. Offenheiser
President, Oxfam America
Daniel Rodriguez
International Technology Development Group (ITDG), Latin America Regional Office, Peru

UN BODIES

Ivar Baste
Chief, Environment Assessment Branch, UN Environment Programme
Wim van Eck
Senior Advisor, Sustainable Development and Healthy Environments, World Health Organization
Joke Waller-Hunter
Executive Secretary, UN Framework Convention on Climate Change
Hamdallah Zedan
Executive Secretary, UN Convention on Biological Diversity

AT-LARGE SCIENTISTS

Adrienne Clarke
Laureate Professor, School of Botany, University of Melbourne, Australia
Denis Lucey
Professor of Food Economics, Dept. of Food Business & Development, University College Cork, Ireland, and Vice-President NATURA
Vo-tong Xuan
Rector, Angiang University, Vietnam

PRIVATE SECTOR

Momtaz Faruki Chowdhury
Director, Agribusiness Center for Competitiveness and Enterprise Development, Bangladesh

Sam Dryden
Managing Director, Emergent Genetics
David Evans
Former Head of Research and Technology, Syngenta International
Steve Parry
Sustainable Agriculture Research and Development Program Leader, Unilever
Mumeka M. Wright
Director, Bimzi Ltd., Zambia

CONSUMER GROUPS
Michael Hansen
Consumers International
Greg Jaffe
Director, Biotechnology Project, Center for Science in the Public Interest
Samuel Ochieng
Chief Executive, Consumer Information Network

PRODUCER GROUPS
Mercy Karanja
Chief Executive Officer, Kenya National Farmers' Union
Prabha Mahale
World Board, International Federation Organic Agriculture Movements (IFOAM)
Tsakani Ngomane
Director Agricultural Extension Services, Department of Agriculture, Limpopo Province, Republic of South Africa
Armando Paredes
Presidente, Consejo Nacional Agropecuario (CNA)

SCIENTIFIC ORGANIZATIONS
Jorge Ardila Vásquez
Director Area of Technology and Innovation, Inter-American Institute for Cooperation on Agriculture (IICA)
Samuel Bruce-Oliver
NARS Senior Fellow, Global Forum for Agricultural Research Secretariat

Adel El-Beltagy
Chair, Center Directors Committee, Consultative Group on International Agricultural Research (CGIAR)
Carl Greenidge
Director, Center for Rural and Technical Cooperation, Netherlands
Mohamed Hassan
Executive Director, Third World Academy of Sciences (TWAS)
Mark Holderness
Head Crop and Pest Management, CAB International
Charlotte Johnson-Welch
Public Health and Gender Specialist, International Center for Research on Women (ICRW)
Nata Duvvury
Director Social Conflict and Transformation Team, International Center for Research on Women (ICRW)
Thomas Rosswall
Executive Director, International Council for Science (ICSU)
Judi Wakhungu
Executive Director, African Center for Technology Studies

GOVERNMENTS
Australia
Peter Core
Director, Australian Centre for International Agricultural Research
China
Keming Qian
Director General Inst Agricultural Economics, Dept. of International Cooperation, Chinese Academy of Agricultural Science
Finland
Tiina Huvio
Senior Advisor, Agriculture and Rural Development, Ministry of Foreign Affairs

France
Alain Derevier
*Senior Advisor, Research for
Sustainable Development, Ministry of
Foreign Affairs*
Germany
Hans-Jochen de Haas
*Head, Agricultural and Rural
Development, Federal Ministry of
Economic Cooperation and Development
(BMZ)*
Hungary
Zoltan Bedo
*Director, Agricultural Research
Institute, Hungarian Academy of
Sciences*
Ireland
Aidan O'Driscoll
*Assistant Secretary General,
Department of Agriculture and Food*
Morocco
Hamid Narjisse
Director General, INRA
Russia
Eugenia Serova
*Head, Agrarian Policy Division,
Institute for Economy in Transition*
Uganda
Grace Akello
*Minister of State for Northern Uganda
Rehabilitation*
United Kingdom
Paul Spray
Head of Research, DFID
United States
Rodney Brown
Deputy Under Secretary of Agriculture
Hans Klemm
*Director of the Office of Agriculture,
Biotechnology and Textile Trade Affairs,
Department of State*

FOUNDATIONS AND UNIONS
Susan Sechler
*Senior Advisor on Biotechnology Policy,
Rockefeller Foundation*
Achim Steiner
*Director General, The World
Conservation Union (IUCN)*
Eugene Terry
*Director, African Agricultural
Technology Foundation*

ADVISORY BUREAU

NON-GOVERNMENT REPRESENTATIVES

Consumer Groups
Jaime Delgado
*Asociación Peruana de
Consumidores y Usuarios*
Greg Jaffe
Center for Science in the Public Interest
Catherine Rutivi
Consumers International
Indrani Thuraisingham
*Southeast Asia Council for
Food Security and Trade*
Jose Vargas Niello
Consumers International Chile

International Organizations
Nata Duvvury
*International Center for
Research on Women*
Emile Frison
CGIAR
Mohamed Hassan
Third World Academy of Sciences
Mark Holderness
GFAR
Jeffrey McNeely
World Conservation Union (IUCN)
Dennis Rangi
CAB International
John Stewart
*International Council of Science
(ICSU)*

NGOs

Kevin Akoyi
Vredeseilanden
Hedia Baccar
*Association pour la Protection de
l'Environment de Kairouan*
Benedikt Haerlin
Greenpeace International
Juan Lopez
Friends of the Earth International
Khadouja Mellouli
Women for Sustainable Development
Patrick Mulvaney
Practical Action
Romeo Quihano
Pesticide Action Network
Maryam Rahmaniam
*Center for Sustainable Development
(CENESTA)*
Daniel Rodriguez
*International Technology
Development Group*

Private Sector

Momtaz Chowdhury
*Agrobased Technology and Industry
Development*
Giselle L. D'Almeida
Interface
Eva Maria Erisgen
BASF
Armando Paredes
Consejo Nacional Agropecuario
Steve Parry
Unilever
Harry Swaine
Syngenta (resigned)

Producer Groups

Shoaib Aziz
*Sustainable Agriculture Action
Group of Pakistan*
Philip Kiriro
East African Farmers Federation
Kristie Knoll
Knoll Farms

Prabha Mahale
*International Federation of Organic
Agriculture Movements*
Anita Morales
Apit Tako
Nizam Selim
Pioneer Hatchery

GOVERNMENT REPRESENTATIVES

**Central and West Asia and
North Africa**
Egypt
Ahlam Al Naggar
Iran
Hossein Askari
Kyrgyz Republic
Djamin Akimaliev
Saudi Arabia
Abdu Al Assiri
Taqi Elldeen Adar
Khalid Al Ghamedi
Turkey
Yalcin Kaya
Mesut Keser

**East and South Asia and
the Pacific**
Australia
Simon Hearn
China
Puyun Yang
India
PK Joshi
Japan
Ryuko Inoue
Philippines
William Medrano

Latin America and Caribbean
Brazil
Sebastiao Barbosa
Alexandre Cardoso
Paulo Roberto Galerani

Rubens Nodari
Dominican Republic
Rafael Perez Duvergé
Honduras
Arturo Galo
Roberto Villeda Toledo
Uruguay
Mario Allegri

North America and Europe
Austria
Hedwig Woegerbauer
Canada
Iain MacGillivray
Finland
Marja-Liisa Tapio-Bistrom
France
Michel Dodet
Ireland
Aidan O'Driscoll
Tony Smith
Russia
Eugenia Serova
Sergey Alexanian

United Kingdom
Jim Harvey
David Howlett
John Barret
United States
Christian Foster

Sub-Saharan Africa
Benin
Jean Claude Codjia
Gambia
Sulayman Trawally
Kenya
Evans Mwangi
Mozambique
Alsácia Atanásio
Júlio Mchola
Namibia
Gillian Maggs-Kölling
Senegal
Ibrahim Diouck

TRADE POLICY IMPLICATIONS FOR AFRICA'S AGRICULTURE

Hira Jhamtani

CONTENTS

ABOUT THE AUTHOR

Martin Khor, formerly Director of the Third World Network, is Executive Director of the South Centre, an intergovernmental body with 51 member states from the developing world.

INTRODUCTION

The recent global crisis of high food prices, and of food shortages in some countries, has given prominence once again to food security concerns. In recent years there was complacency about food security and national self-sufficiency, as it was thought that cheaper imports would be always or usually available, and local food production was not so necessary as previously thought. Many developing countries reduced food production, many of them under advice of the international financial institutions.

The rising world prices of many food items in the past few years have meant more expensive imports, and inflation of food prices in local markets. There have also been cases of shortages, as some countries placing orders – for example for rice – have found that the supply is not forthcoming or guaranteed, sometimes because of export restrictions by the exporters of the food item. Many developing countries have been caught in this situation, resulting in street protests as people, particularly the unemployed and poorer families, found it difficult to cope.

Because of this new situation, the paradigm of "food security" has suddenly shifted back to the traditional concept of greater self-sufficiency, instead of prioritizing the option of relying on cheaper imports. It is now recognized that in the immediate future, there is need for emergency food to be supplied to affected countries, but that a long-term solution must include increased local food production in developing countries. This raises the question of what constitute the barriers to local production and how to remove these barriers.

Factors for this crisis include changes in climate (such as drought for example affecting wheat production in Australia), the rising cost of inputs especially oil and oil-based inputs for agriculture, and the switch of land use from production of food to biofuels. However, a longer term reason is the decline in agriculture in many developing countries, in most cases due to the structural adjustment policies of the International Monetary Fund (IMF) and World Bank. The countries were asked or advised to (1) dismantle marketing boards and guaranteed prices for farmers' products; (2) phase out or eliminate subsidies and support such as those for fertilizer, machines, agricultural infrastructure; (3) reduce tariffs on imported food products to low levels.

THE IMPACT OF TRADE LIBERALIZATION

Many countries that were net exporters or self-sufficient in many food crops experienced a decline in local production and a rise in imports which had become cheaper because of the tariff reduction. Some of the imports are from developed countries, which heavily subsidize their food products. The local farmers' produce were subjected to unfair competition, and, in many cases, the farmers could not survive. The effects on farm incomes, on human welfare and on national food production and food security were severe.

Ghana's experience

The case of Ghana[1] illustrates this. The policies of food self-sufficiency and government encouragement of the agriculture sector (through marketing, credit and subsidies for inputs) had supported an expansion of food production (for example in rice, tomato and poultry). The policies were reversed starting from the mid-1980s, and especially in the 1990s. The fertilizer subsidy was eliminated, and its price rose very significantly. The marketing role of the state was phased out. The minimum guaranteed prices for rice and wheat were abolished, as were many state agricultural trading enterprises, and the seed agency responsible for producing and distributing seeds to farmers, and subsidized credit was also ended.

Applied tariffs for most agricultural imports were reduced significantly to the present 20 percent, even though the bound rate (committed in the World Trade Organization,

---

1   See Khor, M. 2008. *The impact of trade liberalization on agriculture in developing countries: The experience of Ghana.* Third World Network, Penang.

WTO) was around 99 percent. This, together with the dismantling of state support, led to local farmers being unable to compete with imports that had become artificially cheapened by high subsidies, especially in rice, tomato and poultry.

- Rice output in the 1970s could meet all the local needs, but by 2002 imports made up 64 percent of domestic supply. Rice output in the Northern region fell from an annual average of 56 000 tonnes (in 1978–80) to only 27 000 tonnes for the whole country in 1983. In 2003, the United States exported 111 000 tonnes of rice to Ghana. In the same year, the United States government gave USD 1.3 billion subsidies for rice. A government study found that 57 percent of rice farms in the United States would not have covered their costs if they did not receive subsidies. In 2000–2003 the average costs of production and milling of white rice was USD 415 per tonne, but it was exported for just USD 274 per tonne, a price 34 percent below its cost of production and processing.

- Tomato was a thriving sector, especially in the Upper East region. As part of a privatization programme, tomato-canning factories were sold off and closed, while tariffs were reduced. This enabled the heavily subsidized European Union (EU) tomato industry to penetrate Ghana, and this displaced livelihoods of tomato farmers and industry employees. Tomato paste imported in Ghana rose from 3 200 tonnes in 1994 to 24 077 tonnes in 2002. Local tomato production has stagnated since 1995. Tomato-based products from Europe have made inroads into other African markets. In 2004, EU aid for processed tomato products was Euro 298 million, and there are many more millions of Euros in indirect aid (export refunds, operational funds for producers' organizations, etc.).

- Ghana's poultry sector started its growth in the late 1950s, reached its prime in the late 1980s and declined steeply in the 1990s. The decline was due to withdrawal of government support and the reduction of tariffs. Poultry imports rose by 144 percent between 1993 and 2003, and a significant share of this was heavily subsidized poultry from Europe. In 2002, 15 EU countries exported 9 010 million tonnes of poultry meat for Euro 928 million, at an average of Euro 809 per tonne. It is estimated that the total subsidy on exported poultry (including export refunds, subsidies for cereals fed to the poultry, etc.) was Euro 254 per tonne. Between 1996 and 2002, EU frozen chicken exports to West Africa rose eight-fold, due mainly to import liberalization. In Ghana, the half million chicken farmers have suffered from this situation. In 1992, domestic farmers supplied 95 percent of Ghana's market, but this share fell to 11 percent in 2001, as imported poultry sells cheaper.

In 2003, Ghana's parliament raised the poultry tariff from 20 percent to 40 percent. This was still much below the bound rate of 99 percent. However, the IMF objected to this move and thus the new approved tariff was not implemented. The IMF representative in Ghana told Christian Aid that the IMF pointed out to the government that the raising of tariffs was not a good idea, and the government reflected on it and agreed. Many farmers' groups and non-governmental organizations (NGOs) in Ghana have protested on this to the government.

THE WORLD TRADE ORGANIZATION AND FREE TRADE AGREEMENTS

Some developments in the trade negotiating arena are also a source of concern. The Doha negotiations at the WTO are mandated to substantially reduce domestic support in developed countries. However, to date, the offers of the United States and the EU indicate their overall trade distorting support (OTDS) would be reduced at the bound level, but not at the applied level. Also, the figures in the Chair's agriculture text of 19 May 2008 would not reduce the actual present domestic support for the United States. The maximum or bound OTDS level for the United States would be USD 13–16.4 billion, while the actual support in 2007 was reported to be around USD 7–8 billion.

Another source of concern is the new Farm Bill of the United States. According to several analyses, including those made by the United States administration, the Farm Bill will continue the present system of subsidies, and in some ways or for several commodities, it will expand the support. For example the Bill guarantees that 85 percent of the domestic market for sugar will be met by local production. The bill also allows a farm family with an income of up to USD 1.5 million to obtain subsidies, compared to the limit of USD 200 000 per farmer proposed by the Bush administration. The Bill thus 'locks in' the United States' system and levels of subsidies for the next five years; it also constrains what its negotiators can offer in the WTO's Doha negotiations.

A major loophole in the WTO's agriculture agreement is that countries are obliged to reduce their bound levels of domestic support that are deemed "trade distorting" but there are no constraints on the amount of subsidies deemed non distorting or minimally distorting, which are placed in the so-called Green Box. Recent studies have shown however that many of the Green Box subsidies are also trade distorting. The Doha negotiations are unlikely to place new effective disciplines on the Green Box. Therefore, the major subsidizing countries can change the type of domestic subsidies they give, while reducing the "trade distorting subsidies" and continue to provide similar levels of farm subsidies.

Meanwhile the developing countries are being asked to reduce their agricultural tariffs further. The Chair's proposal at the Doha talks is for a maximum 36 percent tariff cut for developing countries, and 24 percent for small vulnerable economies. This is sizable, and compares with the 24 percent cut in the Uruguay Round. Most developing countries are advocating that the instruments of special products (SP) and special safeguard mechanism (SSM) be set up as part of the WTO talks to promote food security along with farmers' livelihoods and rural development. SPs would exempt important food products from tariff cuts or at least allow for more lenient cuts. SSM would enable a developing country to impose an additional duty on top of the bound rates in situations of reduced import prices or increased import volume, in order to protect the local farmers. However, there is considerable opposition from some exporting countries to having these instruments work in an effective way.

In the bilateral or regional free trade agreements (FTAs) involving developed and developing countries, the developing countries are asked to reduce or eliminate their tariffs by even more. For example, in the Economic Partnership Agreements (EPAs) between African, Caribbean and Pacific (ACP) countries and the EU, the ACP countries are asked to eliminate their tariffs on 80 percent of their tariff lines over different time periods. Agricultural products are among those affected.

CONCLUSION

- The economic and trade policies followed by many developing countries, often at the advice of international financial institutions, or as part of multi-lateral and bilateral trade agreements, have contributed to the stunting of the agriculture sector in developing countries. The developing countries must be allowed to provide adequate support to their agriculture sector and to have a realistic tariff policy to advance their agriculture, especially since developed countries' subsidies are continuing at a high level. The developed countries should quickly reduce their actual levels of subsidy.

- The agriculture policy paradigm in developing countries must be allowed to change. Countries should have the policy space to expand public expenditure on agriculture. Governments in developing countries must be allowed to provide and expand support to the agriculture sector.

- Developing countries should place high priority on expanding local food production. Accompanying measures and policies should thus be put in place. The countries should be allowed to calibrate their agricultural tariffs in such a way as to ensure that the local products can be competitive, that farmers' livelihoods and incomes are sustained, and that national food security is assured.

- The proposals of developing countries (led by the G33) on special products and special safeguard mechanism, aimed at food security, farmers' livelihoods and rural development, at the WTO should be supported. Effective instruments that can meet the aims should be established.

- The policies of the World Bank, IMF and regional development banks should be reviewed and revised as soon as possible so that they do not continue to be barriers to food security and agricultural development in developing countries.

- The actual levels (and not just the bound levels) of agricultural domestic subsidies in developed countries should be effectively and substantially reduced. There should also be new and effective disciplines on the Green Box subsidies to ensure that this category does not remain an 'escape clause' that allows distorting subsidies that are detrimental to developing countries.

○ There should be a review of many of the FTAs between developed and developing countries, including the EPAs between the EU and ACP countries. In light of the food crisis and the changing paradigm on food security, developing countries that have signed or are in the process of negotiating FTAs should ensure that the FTAs provide enough policy space to allow sufficiently high tariffs on agricultural imports that enable the fulfilment of the principles of food security, farmers' livelihoods and rural development.

INTERVENTIONS IN WATER
TO IMPROVE LIVELIHOODS
IN RURAL AREAS

*Jean-Marc Faurès and
Guido Santini,
in close collaboration with
Rudoph Cleveringa and
Audrey Nepveu de Villemarceau*

CONTENTS

ANNEXES

FIGURES

BOXES

TABLES

ABOUT THE AUTHORS

Jean-Marc Faurès and **Guido Santini** are with the Water Resources, Development and Management Unit of the Food and Agriculture Oraganization of the United Nations (FAO).

Rudoph Cleveringa and **Audrey Nepveu de Villemarceau** are with the Technical Advisory Division – Water Management and Rural Infrastructure of the International Fund for Agricultural Development (IFAD).

EDITORIAL NOTES

This is a reproduction of Chapter 4 of "Water and the rural poor: Interventions for improving livelihoods in sub-Saharan Africa", published by the Food and Agriculture Organization of the United Nations (FAO) in 2008. The editors are grateful for permission to reproduce the material. The figures, tables and boxes in this chapter have been renumbered accordingly.

INTRODUCTION

While water control is often not the only limiting factor in crop production in sub-Saharan Africa (SSA), it is often the starting point for any improvement in agricultural productivity. In many areas, farmers work with poor soils, they have limited financial credit, they apply too little fertilizer, and they are unable to harvest and deliver their crops to market in a timely fashion. However, in many arid and semi-arid regions, the lack of access to water (or inadequate control or timing of water supplies) contributes to the difficulty of generating acceptable yields. In addition, uncertainty regarding rainfall or access to a developed irrigation supply causes farmers to apply less seed and fertilizer than they might otherwise do. Hence, efforts to improve farm-level access to water and control of water deliveries or rainfall will, in the zones described in Annex 1, enable farmers to improve productivity within current cropping patterns and to consider diversifying their crop choices, thus progressively increasing the proportion of their marketable surplus, albeit locally.

Investments and policies that influence how farmers use water in crop and livestock production must be evaluated according to the local conditions in order to ensure that policy guidelines and parameter values address poverty reduction goals effectively. Opportunities to reduce poverty by improving access to water and the types of investments that will be most helpful in increasing agricultural productivity and improving rural livelihoods will vary among regions according to the prevalence of rural, subsistence farming, the types of livelihood zones, agro-ecological zones and climate. So too, will the types of investments and associated institutional measures needed to achieve poverty reduction goals. Decisions regarding water development for agriculture must consider both biophysical and socio-economic aspects of water resource availability and management.

The analysis of poverty patterns in SSA and their links to agricultural practices calls for specific attention to the improvement of rainfed agriculture. In all such areas, intervention programmes must address as a priority the needs of poor smallholder farmers located far from markets and those who lack secure water rights. Some of these rainfed areas could benefit from investments in new, large-scale irrigation infrastructure (especially where better-off producers have access to markets and less well-endowed people can find decent employment in upstream or downstream activities, such as agroprocessing) (FAO, 2006a). In other places, livestock production, inland fisheries and aquaculture, or other types of multiple water-use systems, will need to receive special attention.

When working on a national scale, the range of different livelihood realities has to be taken into account. Large differences can exist in a country between one region and another in terms of agricultural practices, natural resources endowment (in particular soil and water), market opportunities, knowledge and education levels, and the capacity of local institutions. Such differences need to be taken into account in developing water control strategies that match the needs and capacities of local populations. The key term is "context-specificity".

Notwithstanding the differences that are relevant, a key observation is that successful efforts to improve crop yields and farm incomes in SSA will require concerted efforts to intensify crop production on small-scale farms (Abalu and Hassan, 1998). In most cases, when dealing with such farms, investments in improved water control will not be feasible without considering a range of conditions for success. These conditions are discussed below.

MATCHING THE SPECIFIC NEEDS OF DIFFERENT GROUPS

This study has attempted to estimate the relative importance of four main categories of farming populations in SSA (Figure 1). While the estimates are relatively approximate, in most countries of the region, the bulk of the farming population (330 million or about 80 percent) is represented by traditional smallholders, producing mainly staple food for household consumption and with relatively marginal connections to markets. Other major categories include: highly vulnerable people, living at the margin of survival (50 million or 12 percent); emerging

FIGURE 1

A typology of farming populations in sub-Saharan Africa

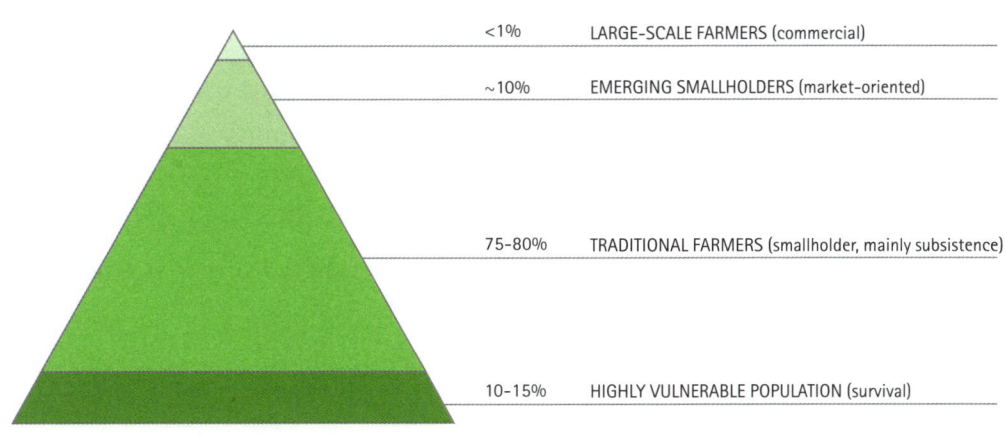

<1%	LARGE-SCALE FARMERS (commercial)
~10%	EMERGING SMALLHOLDERS (market-oriented)
75-80%	TRADITIONAL FARMERS (smallholder, mainly subsistence)
10-15%	HIGHLY VULNERABLE POPULATION (survival)

smallholder farmers, who may partially subsist from their own production but whose principal objective is to produce a marketable surplus (40 million or 9 percent); and commercial farmers and enterprises oriented towards internal and export markets (less than 2 million or 0.5 percent). In addition, it is estimated that the non-agricultural population represents 7 percent of the rural population in SSA (FAOSTAT, 2008). Each of these groups faces different constraints, and each needs adapted responses in all fields, including water control.

Each of these groups has to be addressed in a different way, as shown in Figure 2. In most cases, the highly vulnerable populations in rural SSA consist of people having no or very limited access to land and other livelihood assets. They are often landless workers, widows, families affected by HIV/AIDS or other diseases, etc. For these people, water interventions should focus on highly subsidized social programmes, including labour-intensive soil and water conservation or watershed management programmes that can provide a return on labour. Domestic water supply and sanitation programmes also have good potential for impact, in part through reduction in water-related diseases and in time spent for fetching water.

The smallholder farmers in rural SSA require investments in rainfed water management and supplementary irrigation where feasible. They need secure land tenure that is stable and reliable, guaranteed access to water, support to the empowerment of local communities, in particular water users associations (WUAs), and improved access to inputs (through targeted subsidies) and markets. Capacity building, education and agricultural extension are also important, in addition to domestic water and sanitation programmes. Helpful public interventions will include research and development and extension support for maximizing yields with limited resources, diversifying crop production alternatives and producing more than one crop per year, where feasible.

Compared with traditional smallholders, emerging farmers typically have a higher level of technical knowledge and are more receptive to improved technology. They tend to specialize in specific crops, and are often integrated into a production/supply chain with some support from buyers through extension services and input supply. As they progress in market-oriented production, emerging farmers increasingly need to better secure production inputs. Together with fertilizers, improved control of soil moisture through irrigation is an important element of their production strategy. Therefore, access and control of water are essential, together with improved access to well-adapted financial instruments.

A subcategory of emerging farmers comprises those who produce crops on very small plots of land in home gardens or on other small landholdings, close to local markets. Small-plot irrigation technologies include treadle pump, affordable drip irrigation kits and water storage options (Keller and Roberts, 2004). These technologies are characterized by low initial investment costs, relatively short payback periods, and high farm-level returns on investments (Magistro et al., 2007). In addition, widespread use of small-plot irrigation methods can generate employment opportunities on and off farms in rural areas. Treadle pumps and drip systems are somewhat labour-intensive, and local entrepreneurs can establish businesses that build, service and repair the irrigation equipment. Such activities stimulate greater demand for farm products and other non-tradable goods and services.

PLEASE MAKE SURE THAT YOUR PRESCRIPTION IS... *(faded)*

and has been **marked SLS on an FP10 prescription** and includes the product name, order code and PIP code. Please sign the prescription on the reverse side and include your date of birth. Please pay the prescription charge if applicable. You are exempt from the charge if you are over 60 or have a specified medical condition and have a valid medical exemption certificate (MedEx) for example.

Note that it is possible to order a **PACK OF THREE RINGS ALL IN THE SAME SIZE** by specifying that size on the prescription. **Pharmacist MUST relay sizing requirement to iMEDicare directly.**

NHS Order Code:

Assist Maintenance System - (PIP code: 3131232)
(£27.00) (Loading cone, applicator collar, 3 Select rings and lubricant).

S M L

☐ 15109

High Tension

Ultra Maintenance Ring Set - (PIP code: 317-5619).
(£34.00) (3 sizes High tension - venous leak and narrow penile diameter).

(SIZES: 4 5 6)

☐ 15444

Ring Tension

Ultimate (SUREfit) Maintenance Ring Set (PIP code: 317-5601)
(£34.00) (3 sizes medium-high tension - super comfort - known also as SureFit).

(SIZES: ~~X~~ [Y] ~~Z~~)

☑ 15222

Select Maintenance Ring Set (PIP code: 3131661).
(£29.00) (3 sizes - low to medium tension)

(SIZES: 4 5 6)

☐ 15230

Low Tension

SureEase Maintenance Ring Set - (PIP code: 3131240).
(£32.00) (2 sizes SureEase Comfort - soft, removal loops. 2 sizes SureEase Ultra - durable).

(SIZES: 2 2 3 4)

☐ 15085

Yours sincerely

SIGNED...NAME (Capitals)..

TITLE..or circle: Consultant, Associate Specialist, Registrar, Staff Grade, Specialist Nurse Practitioner, Clinical Assistant, SHO.

Directorate...

iMEDicare LTD, Unit 11, Shakespeare Industrial Estate, Shakespeare Street, Watford WD24 5RR +44 1923 23 77 95, contact@imedicare.co.uk

> **Schedule 11 Prescription Criteria for Vacuum Therapy** - Diabetes - Multiple Sclerosis - Parkinson's Disease - poliomyelitis - prostate cancer - prostatectomy - radical pelvic surgery - severe pelvic injury - renal failure treated by dialysis or transplant - single gene neurological disease - spinal cord injury - spina bifida.

○ Continue to repeat protocol above minimum 2-3 times / week there-after to maintain good penile health, further improve penile circulation, and optimize chance of erectile recovery, where possible.

Erection Maintenance Ring Set - Types:

		Large	Medium	Small	NHS Order Code:
High Tension -	**Ultra**	6	5	4	**15444**
Medium Tension -	**Ultimate Surefit**	Z	Y	X	**15222**
High Tension -	Select	6	5	4	**15230**
PRACTICE RINGs -	**SureEase**	4	3	Two of size 2	**15085**

Two Rings Simultaneously? (Each ring applied separately with additional pumping before application of 2nd ring).

Most likely not necessary_____ Perhaps _____ Definitely yes _____

Curvature Correction? (Peyronie's Disease): Yes / No

Protocol: Generate 20 full erections in sequence holding each FULL erection in the correct diameter cylinder for a maximum of 5 seconds. To be repeated daily for a minimum of 6 months until curvature correction optimized (maximum 12 months). **Cylinder sizing option for curvature correction**: (alternate daily between)

XLL (Extra Large Long) _____ XL (Extra Large)_____ C (Large)_____ C & B (Medium)_____ C & B & A (Standard)_____

Contact for further assistance Call +447825226468 or email: norththamesrep@imedicare.co.uk

Order **SOMATherapy-ED 140g** Sealing Lubricant (or **"YES WB"**: vaginal dryness) by mail order on **01923 23 77 95 or** via **www.iMEDicare.co.uk**. Do not use vaseline or baby-oil with maintenance rings - use waterbased gel only.

Text

Name: Shaz

K Y Jelly

iMEDicare SOMAerect Demonstration Summary sheet:

Device / Model recommended:

SOMACorrect Xtra - (PIP code: 317-5593).

SOMACorrect (Peyronie's Only) - (PIP code: 313-1653).

SOMAerect Response II XL / (PIP code: 317-5585)

SOMAerect Response II (PIP code: 3131182)

SOMAerect Touch II - (PIP code: 3131190)

NHS Order Code

15111

15080

15888 (XLL cylinder from iMEDicare only - not on NHS)

15019

15013

Cylinder sizing option(s): (Black inner - outer / grey outer) cushioning inserts)_____

XLL: (<u>Extra Large Long</u> Grey Outer cushioning insert)_____

XL: (<u>Extra Large</u> Grey Outer cushioning insert)_____

C: (<u>Large</u>   Black Outer cushioning insert)_____

C & B: (<u>Medium</u> - Black Outer cushioning insert) _____

C & B & A: (<u>Standard</u> - Black Inner and Outer cushioning inserts) _____

Pump Slowly!

Rings last 30-40 uses approx

Max 2 rings at a time

Never use has rings if on blood
thinners stronger than Aspirin 75mg

Pumping Protocol: 2-3 pumps every 5 / 10 seconds_____   Trim / Shave pubic hair_____

Hold scrotal skin during pumping? (helps to minimize loose tissue entrapment at open end of cylinder) Yes / No

INITIAL RECOVERY PHASE Protocol / Physiotherapy (ED):

10 to 20 full erections (as full as you can tolerate initially) in a row holding each at full erection for not more than 5 seconds, **WITHOUT USE of Maintenance Rings** - reverses penile shrinkage and improves pliability to restore normal comfort at full erection, improves penile blood flow and toughens weakened tissues. Trim or shave pubic hair.

iMEDicare Replacement Rings Prescription Advice form

Hospital: ....

Dear Doctor: .... **Date:** ....

Regarding patient: ....

Surname: Hospital No.:
First names(s):   DOB:
Address:
SPACE FOR PATIENT IDENTIFICATION LABEL

DELIVERY:

Patients are requested to take their prescription to their local community pharmacy who should order directly on

T: 01923 23 77 95
F: 01923 804 206
E: contact@imedicare.co.uk

or from the "Specials Depts"
of Wholesalers
FREE delivery to the Pharmacy
from iMEDicare Ltd.

Your patient was seen in the outpatient department at this Hospital today. Following physical assessment and full instruction in correct usage it has been established that the device / ring options recommended below will prove most suitable and effective for him as a solution to his erectile dysfunction. Can you please prescribe:

ERECTILE DYSFUNCTION MANAGEMENT SYSTEMS

(Vacuum Therapy Devices or Accessories) **To GP** - Please mark prescription 'SLS' (Selected List Scheme)

How to obtain your Replacement Rings on prescription from your pharmacy:

You can take your fully completed prescription to any Local Community Pharmacy - ask them to order directly on the **iMEDicare order line - 01923 23 77 95** for a FREE delivery

FIGURE 2

Adapting agricultural support strategies to different farmers groups

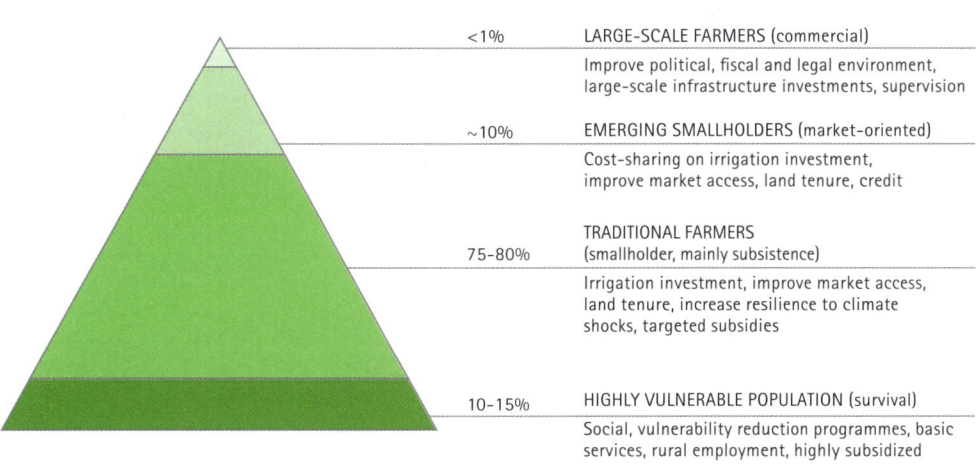

Finally, there are the commercial farmers. Their activities usually offer local development opportunities, in particular for landless workers, and contribute to local economies. Therefore, commercial farming should be considered as a potentially important element in rural poverty reduction programmes, alongside programmes that address the needs of other categories. Commercial farmers typically benefit from favourable political, institutional and fiscal environments, good transportation, storage and marketing infrastructure, and reductions in international trade barriers. They are also well equipped to enhance the profitability of large-scale irrigation infrastructure. Where provided with the right legal framework, and when a fair and transparent balance of power is guaranteed, commercial and emerging farmers can benefit the rural poor through fair, decent and gainful employment options and, thus, contribute to local poverty reduction.

Beyond the broad categories of farmers described above, a further and more refined distinction between target groups needs to distinguish between farmers, herders, fishers, and landless and migrant labourers. Gender specificities must be taken into account through a differentiated needs analysis for men, women, children, young and elderly people. Here, the livelihood concept provides a valid framework that enables an understanding of the different types of assets they use to sustain to their livelihood, and, therefore, helps in identifying their specific needs in terms of livelihood assets consolidation. The special case of people affected by HIV/AIDS is highly relevant in several SSA countries (Box 1).

BOX 1
HIV /AIDS AND IMPLICATIONS FOR WATER INTERVENTIONS

The rapid progression of the HIV/AIDS pandemic is having a particularly devastating effect on the rural poor, and rural women specifically as their traditional care-giving role makes them bear the burden of looking after the sick and orphans while also securing a livelihood for the household. The loss of labour in HIV/AIDS-affected households and the resulting reduction in the area of land cultivated (resulting in lower production), the shift to less labour-intensive crops and delays in agricultural operations all undermine households' food security status.

HIV/AIDS worsens gender-based differences in access to land and other productive resources such as labour, technology, credit and water. In many cases, legal and customary law do not allow widows to retain access and control over land and water. In other cases, their water rights are not respected, protected or fulfilled.

Therefore, the introduction of appropriate and affordable technologies for safe water supply and sanitation is of the utmost importance. An increase in the demand for water is also caused by the need for water for productive use, but the weakening of people affected by HIV/AIDS must be taken into consideration in project design and the choice of technologies.

Source: FAO (forthcoming)

OPTIONS FOR INTERVENTIONS IN WATER

Improved water control and management for poverty reduction in rural areas includes a range of technical options to support cropping, livestock, forestry, aquaculture, domestic and other productive activities. In cropping, interventions range from on-farm water conservation practices that focus on improving soil water storage in rainfed agriculture to more elaborate types of water control, moving along the continuum from purely rainfed to irrigated agriculture, first as a means of securing production through supplementary irrigation, then allowing for an increase in the cropping intensity, and allowing for diversification of crop production through "full control" irrigation. Such systems are not mutually exclusive, and several of them can find their application in a single livelihood context. Thus, irrigation provides opportunities for the multiple use of water, including for domestic consumption, aquaculture and livestock within the production system (Molden, 2007). Figure 3 presents a typology of some of the most widespread agricultural water management options.

Based on the above typology, it is possible to establish a list of water-related interventions. Table 1 is adapted from a matrix developed in the framework of FAO's Special Programme of Food Security (FAO, 1998), and shows options for water control by type of use and available technologies, organized along four main water management components: capture, storage, lifting and application. Well adapted to smallholders, who are the main target beneficiaries of the Programme, Table 1 shows the range of possible options to be used as part of poverty reduction strategies in rural areas. A selection of the most relevant options is discussed in more detail below.

Geographical scales offer another way to classify water intervention options. They have significant operational implications, as changes in scale imply changes in approaches and social organization. Plot-level or farm-level interventions, through improved soil moisture management in both rainfed and irrigated agriculture, will rely primarily on farmers' capacity and willingness to adopt improved practices.

At the scale of irrigation schemes, water distribution and management require a higher level of organization, implying the need for effective local water management institutions. Water conservation in small watersheds typically involves several communities along the river, with many social groups having different interests. The level of social organization and institutions needed to address water management adequately increases with the scale of the watershed. Transboundary rivers are the ultimate level of complexity for water management, where political dimensions add to local management issues. While all scales of intervention are important, this study focuses primarily on local-level interventions.

FIGURE 3

A typology of agricultural water management practices showing the diversity of options

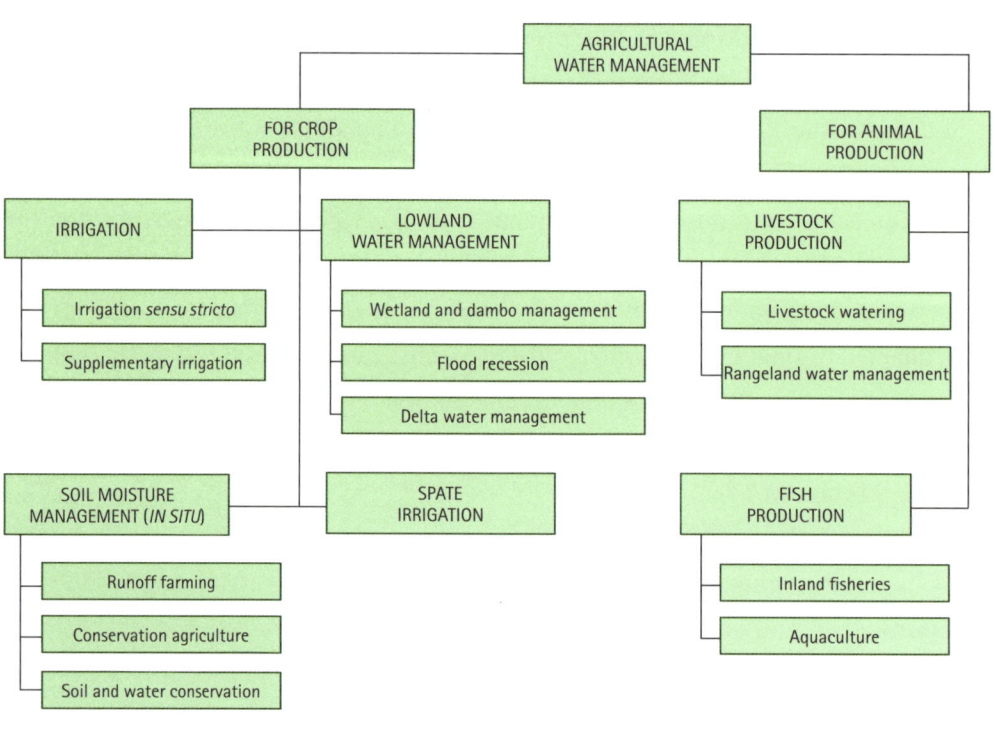

Source: Adapted from FAO–AQUASTAT (2008)

TABLE 1

Indicative list of water control and water use technologies

USES	TECHNOLOGIES			
	Water capture	Water storage	Water lifting	Water use/application
Domestic water use (safe drinking-water for cooking, bathing, laundry, cleaning)	Shallow tubewells: ○ dug wells ○ drilled wells Spring diversion Deep tubewells Recharge enhancement system: ○ recharge wells Underground water harvesting system: ○ cistern or other underground water storage structure fed by a catchment area Above ground rainwater harvesting system: ○ rooftop tank or jar		Human powered pumps: ○ hand pulleys and buckets ○ hand pumps Solar pumps Motorpumps	Water purification methods: ○ filters (e.g. sand filters) ○ boilers for drinking-water ○ chlorination
Irrigated crops (including urban and small plot cropping)	Shallow tubewells: ○ dug wells ○ drilled wells Spring diversion Deep tubewells Water harvesting systems, composed of: ○ catchment area and a water storage structure aboveground (e.g. excavated pond, impounded reservoir) ○ catchment area and a water storage structure belowground (e.g. cistern)	Elevated tanks/drums	Human-powered pumps: ○ hand pulleys and buckets ○ hand pumps ○ treadle pumps Animal-powered pumps: ○ mohte ○ Persian wheel Motorpumps: ○ petrol ○ diesel	Aboveground: ○ shallow trenches or ditches ○ family/drum drip irrigation kit ○ low cost hose irrigation system Belowground: ○ porous ceramic jars ○ porous and sectioned pipe
Supplementary irrigation	Shallow tubewells: ○ dug wells ○ drilled wells Deep tubewells Run off the river diversion Water harvesting systems composed of: ○ catchment area and a water storage structure above ground (e.g. excavated pond, impounded reservoir) ○ catchment area and a water storage structure below ground (e.g. cistern)	Small dams/reservoirs	Human-powered pumps: ○ hand pulleys and buckets ○ hand pumps ○ treadle pumps Animal-powered pumps: ○ mohte ○ Persian wheel Motorpumps: ○ petrol ○ diesel	
Enhanced water management for rainfed	Soil and water conservation and management (runoff farming): ○ stone bunds, ridges, broad beds, furrows ○ no-tillage ○ infiltration pits ○ contour bunds (semi-circular, triangular) ○ vegetative bunds ○ terraces (eyebrow, Negarim) ○ mulching			
Aquaculture and inland fisheries	Run off the river diversion	Small dams and reservoirs Integrated paddy and fish production		Basins Ponds Water-level control in small streams

USES	TECHNOLOGIES			
	Water capture	Water storage	Water lifting	Water use/ application
Livestock watering	Shallow tubewells: ○ dug wells ○ drilled wells Spring diversion		Human-powered pumps: ○ treadle pumps Animal-powered pumps: ○ mohte ○ Persian wheel Motorpumps: ○ petrol ○ diesel	Watering facilities: ○ watering troughs
	Water harvesting systems composed of: ○ catchment area and a water storage structure above ground (e.g. excavated pond, impounded reservoir) ○ catchment area and a water storage structure below ground (e.g. cistern)			
	Micro-catchment water harvesting systems for rainwater runoff: ○ contour bunds (semi-circular, triangular)			

Source: Adapted from FAO (1998)

Managing soil moisture at field level in rainfed areas

A key challenge in SSA is to reduce water-related risks posed by high rainfall variability in the semi-arid areas (Rockström *et al.*, 2007). In most areas dominated by rainfed agriculture, there is generally enough rainfall for good yields in rainfed cropping, but it is available at the wrong time and at too great an intensity, followed by dry spells. As a result, most of the rain is lost in unproductive evaporation or surface runoff that causes erosion and loss of soil fertility.

In such areas, investments are needed to assist farmers to establish better control and management of intermittent water supplies (Rockström, 2000; Mupangwa, Love and Twomlow, 2006). These investments should be accompanied by technical assistance for optimizing the use of fertilizer, seeds and other key inputs in rainfed settings when soil moisture management practices are developed. Farmers' risk-aversion strategies, which include low levels of investment in rainfed cropping, can only be modified if their perception of water-related risks changes as a result of such investments.

Especially important in designing soil moisture management investments is distinguishing between droughts and dry spells. In semi-arid and dry subhumid livelihood zones, rainfall variability generates dry spells (short periods of water stress during critical growth stages) almost every rainy season (Barron *et al.*, 2003). In contrast, droughts are major reductions in the amount of rainfall, and they occur on average only once or twice every decade in semi-arid regions. While investments in water management can help mitigate the effects of dry spells on crop yields, droughts cannot be bridged through agricultural water management. Instead, they require institutional and social coping strategies, such as cereal banks, insurance schemes and relief food distribution. The range of differences between dry spells and droughts is given in Table 2.

TABLE 2

Types of water stress and underlying causes in semi-arid and dry subhumid tropical environments

	DRY SPELL	DROUGHT
Meteorological		
Frequency	Two out of three years	One out of ten years
Impact	Yield reduction	Complete crop failure
Cause	Rainfall deficit of two- to five-week periods during crop growth	Seasonal rainfall below minimum seasonal plant water requirement
Agricultural		
Frequency	More than two out of three years	One out of ten years
Impact	Yield reduction or complete crop failure	Complete crop failure
Cause	Low plant water availability and poor plant water uptake capacity	Poor rainfall partitioning, leading to seasonal soil moisture deficit for producing harvest (where poor partitioning refers to a high proportion of runoff and nonproductive evaporation relative to soil water infiltration at the surface)

Source: Rockström *et al.* (2007)

Field-level soil moisture management practices encompass a large range of agronomic practices aimed at better capturing and maintaining water in the rootzone. They include soil and water conservation and "run-off farming" practices (methods aimed at capturing water as it falls on the plot, so as to increase its infiltration rate and reduce runoff). Runoff farming techniques are gaining increasing attention in areas such as western Sudan, where results are very encouraging for improving agricultural production and livelihoods (semi-desert to semi-arid climates). Farmers have obtained significantly improved results when combining traditional moisture control techniques with soil fertility management practices within existing cereal-based livelihood zones. For example, for sorghum production in Mali, Burkina Faso, Niger, etc., improved zai/tassa planting pits catch more of the sparse rainfall, and dung/compost added to the pits enables more efficient use of plant nutrients and moisture. Box 2 gives an example of soil moisture management for rainfed rice production.

The most promising prospect for on-farm moisture management appears to be the various types of conservation agriculture practices that have been developed primarily in Latin America and are now spreading in the SSA context (World Bank, 2007a). Conservation agriculture practices aim to enhance the quality of the soil through practices that reduce, change or eliminate tillage and avoid the burning or exportation of residues (FAO, 2001). Conservation agriculture favours the building up of organic matter in the soil, thus increasing its moisture holding capacity. Conservation agriculture illustrates the interlinkage between soil moisture and soil fertility, and the importance of addressing both issues simultaneously in cropping improvement programmes (Box 3).

A shift from conventional to conservation agriculture requires a package of interventions, including changes in technology (sowing, and weed control), supported by information and training (FAO, 2005). Benefits from conservation agriculture take time to appear, and programmes to promote it among farmers need to be developed with a medium-term perspective. Farmers may need financial support, or assistance in kind, in order to adopt conservation agriculture practices.

BOX 2
SOIL MOISTURE MANAGEMENT FOR RAINFED RICE PRODUCTION

There is substantial opportunity for enhancing rice production and farm incomes in West Africa and the Sahel by improving farm-level access to irrigation water and improving water management in rainfed conditions, in conjunction with other agronomic and crop management improvements. Researchers at the West Africa Rice Development Association (WARDA) and others have demonstrated significant differences between the rice yields obtained on farms and experiment stations (Haefele *et al.*, 2001; Wopereis-Pura *et al.*, 2002; Poussin *et al.*, 2003). Much of the observed yield gap is a consequence of suboptimal weeding strategies and inappropriate use of nutrients (Haefele *et al.*, 2000). However, yields can also be increased by constructing bunds and canals to improve water management in rainfed conditions (Sakurai, 2006). Extension agents can encourage farmers in the region to implement such measures by demonstrating the risk-reducing characteristics of soil and water conservation efforts (Baïdu-Forson, 1999).

BOX 3
CONSERVATION AGRICULTURE IN SUB-SAHARAN AFRICA

Conservation agriculture has started to spread in Africa, and it is being adopted in most subhumid regions. Some farmers have doubled or even tripled their grain yields. In Kenya and the United Republic of Tanzania, FAO is implementing a conservation agriculture project with small-scale farmers in eight districts. In Zambia, conservation agriculture has helped vulnerable households survive drought and livestock epidemics, and more than 200 000 farmers are now using this technique. In the 2000–01 drought, farmers who used conservation agriculture managed to harvest one crop, others farming with conventional methods faced total crop failure. In Ghana, more than 350 000 farmers now use conservation agriculture.

Subsidies to support adoption of conservation agriculture programmes often find additional justification in the environmental benefits they typically provide at the watershed level.

Rainfed moisture management practices find their application mostly in cereal-based and highland temperate livelihood zones, where rainfall ranges between 500 and 2 000 mm. In more arid areas, e.g. agropastoral zones, they face the double challenge of excessive occurrence of dry periods and competition for scarce biomass for different uses, in particular livestock.

Investing in small-scale water harvesting infrastructure

Water harvesting encompasses any practice that collects and stores runoff for productive purposes (FAO, 1994). It includes three components: a watershed area to produce runoff, a storage facility, and a target area for beneficial use of the water (agriculture, domestic or industry). For the purposes of this study, water harvesting is primarily concerned with the construction of small reservoirs, which can serve different purposes (e.g. supplementary irrigation, livestock watering or fisheries, and aquaculture). Different water harvesting systems can be classified according to the scale of runoff collection, from small check dams and water retention structures to larger external systems collecting runoff from watersheds (Oweis, Prinz and Hachum, 2001). Storage options in rainwater harvesting include surface or subsurface tanks and small dams (Fox and Rockström, 2000).

Water harvesting techniques are used in a range of contexts in drylands to concentrate and make more effective use of rainwater, and to enhance the reliability of agricultural production. However, they are restricted to specific environmental and socio-economic conditions. There is no clear distinction between *in situ* soil water control and management and water harvesting, and several authors refer to a continuum of water management practices from rainfed to irrigated agriculture.

The potential for poverty reduction through water harvesting is high in smallholder settings in semi-arid and subhumid areas. Investments in small reservoirs (typically providing 1 000 m³ of extra water per hectare per season) for supplementary irrigation improve farmers' resilience to dry spells, and, in combination with improved soil, nutrient and crop management can substantially increase the productivity of small-scale rainfed agriculture (Rockström *et al.*, 2007).

Water harvesting technologies have been successfully developed over many years by populations seeking to improve water control. Many ancient water harvesting practices are today widely applied and adapted, such as "half-moons" in West Africa. Others have tended to be abandoned, as economies develop and labour costs of maintenance become excessively high. However, there is still scope for better dissemination of a range of water harvesting technologies that are still relatively little known outside their area of origin. Box 4 provides an example of the range of water conservation options that can be adopted in a semi-arid environment.

All new or adapted water harvesting technologies need to take local socio-economic aspects adequately into account. Labour-saving devices are particularly relevant in areas where labour is scarce or losing its work potential, as is the case with people affected by HIV/AIDS in stricken regions of Africa and Asia. Cultural and socio-economic knowledge and an excellent capacity for understanding and exchanging with farmers are fundamental to the sharing of concepts and practices.

A range of successful water harvesting examples show promise for climate change adaptation: reducing the risks of crop production (including trees) associated with high rainfall variability in semi-arid regions; reducing wind erosion; enhancing aquifer recharge; and allowing for careful expansion to areas where rainfall is normally not sufficient.

Improved ploughing techniques have proved effective for large-scale operations for reclaiming degraded lands. Two ploughs, the "Delfino" (dolphin) and the "Treno" (train) adapted to different soil types are able to reclaim large areas of degraded land through creating "half-moon" microbasins for water capture. This technology, which has been tested in ten countries (Burkina Faso, Chad, Egypt, Kenya, Morocco, Niger, Senegal, Sudan, Syrian Arab Republic and Tunisia), has potential for extensive land reclamation in the most arid areas of the region. However, it is a highly mechanized technique and, therefore, suitable primarily in areas where labour is scarce.

Water harvesting techniques are most relevant in semi-arid and subhumid zones, in particular in cereal-based, agropastoral and Southern African smallholder zones where water is needed in order to supplement rainfall during dry spells.

BOX 4
THE KEITA PROJECT: EXPLORING OF THE RANGE OF WATER CONSERVATION OPTIONS IN WESTERN NIGER

The Keita Project, funded by Italy and the World Food Programme, started its activities in the Ader-Doutchi-Majiya, an arid region of Niger, in 1984. It is a project of unusual scale and duration, and by 1991, it covered an area of 13 000 km², with about 300 000 people in 400 villages. The project provided services and infrastructure on a grand scale. By the end of 1999, it had created 50 artificial lakes, 42 dams and 20 anti-erosion dykes, and 65 village wells. It had applied soil and water conservation techniques to about 10 000 ha of land, and had planted 16 million reforestation seedlings. In addition, the project had built a series of infrastructures, including schools, maternity centres, veterinary facilities, shops, and storehouses, and it included women's empowerment programmes, microcredit, and adult literacy courses. The aspects of the project that were most appreciated by the local population were the increased availability of water and fodder, together with the distribution of "food for work" in an area with few work opportunities (Rossi, 2005). Ten years after project completion, most of the hydraulic infrastructure was still in place and functioning for the benefit of local populations (FAO, 2002).

Promoting community-based small-scale irrigation

While large public investments in irrigation imply a concentration of production factors in a few selected locations, small-scale water control facilities have the potential to affect poverty reduction at local level, contributing to the development of local markets and rural economies. However, experience has shown that a series of conditions need to exist in order to guarantee the success of such irrigation schemes.

Social cohesion and the absence of political interference are a first condition for the success of small-scale irrigation systems. Too often, the relatively high cost of irrigation investments attracts the attention of local politicians, leading to exploitation by clientele and patronage systems. Where associated with the absence of a strong community governance capacity, such conditions lead to inappropriate decisions, inequity in access to irrigated land, and the rapid degradation of infrastructure owing to a lack of maintenance.

In most cases, the design of small-scale irrigation systems holds the key to their sustainability. Operational simplicity is among the most important criteria for the success of small-scale community-based irrigation schemes. The number of users sharing a common infrastructure should remain low, and be based on existing social constructs. Such systems must also be robust, with low maintenance requirements, and limited physical and financial capital requirements – all factors contributing to an easier appropriation of the technology by the users. The planning and design of small-scale irrigation schemes must also give greater attention to water resources and ensure that the schemes will be provided with adequate water supply throughout the cropping season.

Community participation in the design and realization of small-scale irrigation schemes is the only way to ensure beneficiary appropriation, which in turn will facilitate the sustainable management of the investments (Boxes 5 and 6). In the past, too many irrigation systems were designed without considering people s requirements and management considerations. The result was blueprint designs that were not adapted to local conditions, unnecessarily high operation and maintenance costs, and complex organizational settings.

BOX 5
SMALL-SCALE IRRIGATION IN UGANDA

Many small irrigation projects have been implemented with the goal of reducing poverty in rural areas where agricultural productivity is constrained by inadequate access to water. Successful examples include a community-run water project in Uganda that provides equitable access to valley tanks for harvesting rainwater, and a wind-powered irrigation system that has improved livelihoods in the United Republic of Tanzania. The latter project provides irrigation and a water supplyline for domestic use to the centre of a village. Farmers were unable to afford the capital cost of investing in such a programme on their own. The success of the project has inspired eight neighbouring communities to replicate it. In Kenya, the Dryland Development Centre of the United Nations Development Programme (UNDP) links poor people of dry areas in Nairobi with people who have knowledge of key topic areas, such as water management.

Source: IFAD (2005)

Such conditions often imply choosing designs that do not correspond to the lowest-cost investment option, but they do guarantee sustainability in the control of infrastructures by the users. Indeed, while unit costs of small-scale irrigation may not be lower than for large systems, i.e. there are economies of scale (Inocencio *et al.*, 2007), adopting smaller-scale schemes in the framework of larger projects could show higher economic returns and have higher impacts than large systems in terms of poverty reduction in rural areas.

BOX 6
THE POTENTIAL FOR IRRIGATION DEVELOPMENT IN ETHIOPIA

The potential for increasing the irrigated area, and associated agricultural outputs and farm incomes, in Ethiopia is substantial. Godswill, Kelemework and Aredo (2007) compared irrigated and rainfed yields in a study involving about 300 households in three small-scale irrigation schemes in the Rift Valley. They observed mean output values of Br 2 702 per hectare on rainfed farms (average size 1.5 ha) and Br 29 474 per hectare (11 times more) on irrigated farms (average size 0.45 ha). Households with irrigation apply more seed, pesticide, fertilizer and labour than households without irrigation.

In another study, Diao and Pratt (2007) examined the potential economic impacts of expanding irrigated area in Ethiopia using an economy-wide simulation model. They compared an irrigation scenario based on Ethiopia's Irrigation Development Programme, in which irrigated area expands by 274 000 ha by 2015, with a "business as usual" scenario that simply extends the trend in irrigated area observed between 1995 and 2002. The authors concluded that the increase in irrigated area (50 percent of which would be allocated to cereal crop production) would increase the annual economic growth rate from 1.9 to 2.1 percent by 2015. With complementary investments in markets and transportation infrastructure, GDP would increase by 3.6 percent/year.

Small wetlands, dambos and other lowland valley bottoms have always represented a good opportunity for agricultural production, in particular rice, in large areas of SSA, thanks to the availability of water. Wetlands and valley bottoms that have benefited from external investment to improve water control in SSA represent about 555 000 ha and those cultivated directly by farmers without external investments cover about 1 million ha. In addition, flood recession cropping is practiced on another 960 000 ha (FAO, 2006b). Substantial improvements can be made through the introduction of simple technologies in lowlands, including small dams, pumps or affordable well digging. Investments can enable farmers to make better use of lowland areas near urban centres, such as planting two crops of rice per year (Erenstein, Oswald and Mahaman, 2006).

With appropriate policies in place and incentives to local producers, investments in small-scale irrigation could maximize the value of recent developments in rice breeding. The "new rice for Africa", known also as NERICA, generates substantially higher yields per hectare than traditional varieties, but it requires optimal control of soil moisture and nutrient conditions (Dalton and Guei, 2003; Kijima, Sserunkuuma and Otsuka, 2006). Higher rice yields have the potential to improve farm incomes, to increase the aggregate supply of rice in the region, and to limit rice imports at regional level. It is as a result of exploiting these advantages that rice consumption as a proportion of cereal consumption increased from 14 percent in 1970 to about 25 percent in 1990 (Otsuka and Kalirajan, 2006).

While small-scale community-based irrigation systems are valid options in almost all types of livelihood zones, they are most relevant in areas where water is a constraint on crop production, i.e. in semi-arid to subhumid zones.

Improving existing irrigation systems

Irrigation projects in SSA, in particular large-scale projects, have a reputation of high cost and low sustainability. Although there were many failures in the 1970s and 1980s, more recent projects have generally had acceptable rates of return (World Bank, 2007a). Key factors associated with higher rates of return to irrigation development in SSA include lower per hectare costs, market access, and production systems that use inputs more intensively – the last two being strongly correlated. However, irrigation projects continue to have a mixed track record on sustainability. The frequent need for rehabilitation projects in both large- and small-scale irrigation in SSA (Sudan, Madagascar and Mali) shows the poor sustainability of investments in the sector, and the rates of return of externally financed projects have sometimes had to be revised downwards. Today, about 25 percent of the 7.1 million ha of land equipped for irrigation in SSA are out of use for one reason or another (FAO-AQUASTAT, 2008).

The reasons behind poor performance of existing irrigation schemes have been studied extensively (Aw and Diemer, 2005; Morardet *et al.*, 2005). They vary from technical and economic to institutional and social. They include lack of adequate consideration for land tenure and water security issues, overoptimistic hydrological analysis (IFAD, 2005), neglect of water governance and institutional capacity issues, and an absence of adequate environmental assessment. Falling prices of main agricultural commodities, associated with poor evaluation of markets and profitability, and the absence of agricultural support packages were also among major causes for failure. Furthermore, such projects were often characterized by poor and overly complex technical designs, resulting in technology choices and high maintenance costs (Morardet *et al.*, 2005; World Bank, 2007a). Typically, there is a range of fundamental socio-economic changes involved with large-scale irrigation. These are often not sufficiently considered during the planning stage. They include the time needed by social organizations to adapt to technological change, which surpasses by far common development project time frames (Diemer and Huibers, 1996).

While several conditions still limit widespread improvement in the productivity of irrigation schemes, rehabilitation of some of the existing infrastructure offers good possibilities where conducted in conjunction with appropriate changes in design and management. Such changes include, in particular, a much more comprehensive involvement of producers at critical stages in the planning process, and the adoption of a management mechanism that empowers farmers and allows for simpler and more efficient water control. Therefore, modernization approaches need to focus on improved infrastructure and management for increased reliability and flexibility in the service of water.

However, success in increasing the productivity of irrigation systems also depends on a range of other considerations that require careful attention. A clear policy and the appropriate instruments to allow farmers to operate in a conducive environment are necessary preliminary conditions. In the case of rice, a fiscal policy that promotes local or regional production is fundamental. Good market linkages, training packages, strengthening of producers' organizations, and well-targeted credit and finance products are key to the success of large-scale irrigated agriculture.

Improving water control for peri-urban producers

Rapid urbanization in Africa provides increasing opportunities for farmers to produce and market crops in peri-urban areas (Drechsel and Varma, 2007). This dynamic sector of activities is often undervalued. Although estimates of existing irrigation activities around cities are unreliable and incomplete, some data indicate that the scale of the activities is large. For example, the area of the 22 formal irrigation schemes in central Ghana is 8 587 ha, while the estimated area of informal irrigation near cities in the same region is estimated at 40 000 ha (Drechsel *et al.*, 2006). In the United Republic of Tanzania, it is estimated that 90 percent of households in representative villages have small plots under informal irrigation.

Informal irrigation around cities grows as a response to good market opportunities. Typically, it is a flexible and demand-responsive production system, mostly run by small-scale farmers producing vegetables and other non-staples (Drechsel *et al.*, 2006). These farmers typically face acute problems of land tenure and access to quality water. Localized sources of water, which include groundwater, streams, urban drains piped water and wastewater, are often heavily contaminated owing to the rudimentary sanitation arrangements and unregulated effluent discharge (Box 7).

BOX 7
SMALL PLOT IRRIGATION

Small plot irrigation or gardening typically ranges from a few square metres to 0.5 ha. It allows single families to produce food for domestic consumption and for the local market, and requires a shallow source of water. For example, treadle pumps and low-cost drip systems can enable farmers to utilize shallow groundwater in some of the 7.5 million ha of dambo wetlands found in Southern Africa (Roberts, 1988). Small-plot irrigation can also reduce women's workloads, create opportunities for women to learn new skills, and reduce the need for family members to migrate away from home in search of seasonal wage labour (Magistro *et al.*, 2007).

Potential capital investments in water control to support small peri-urban farmers range from small check dams and affordable groundwater drilling and casing technologies to small pumps and localized garden irrigation kits. Small irrigation schemes that benefit a small number of producers have also proved successful. They need to be designed for ease of operation and low maintenance costs so that producers groups can manage them easily.

There is probably no other type of investment that requires a more integrated approach than that of peri-urban farming. Paramount to the success of peri-urban agriculture are the successes obtained in securing access to land and water, providing extension in support to diversification, and ensuring the control of health-related hazards.

Investments to support small-scale peri-urban farming are valid across the whole region, and are relevant in all climate conditions. Examples of successful peri-urban horticulture projects range from ones in Kenya (Box 8) to others in Cape Verde and the Democratic Republic of Congo.

BOX 8
URBAN HORTICULTURE IN KENYA

In Kenya, the horticulture industry has expanded substantially in peri-urban areas in recent years. Much of the new production takes place on small-scale, irrigated farms. In areas near Nairobi, sprinkler, drip and furrow systems are used on farms ranging in size from 0.1 to about 1.0 ha. Kulecho and Weatherhead (2006) interviewed a sample of small-scale farmers to determine major issues regarding irrigation of vegetables, particularly with low-cost drip systems. The three problems mentioned by most farmers were: lack of adequate technical support when using the low-cost drip kits; inadequate water supply; and the lack of marketing opportunities for the vegetables produced. These results demonstrate that small-scale farmers need adequate technical support, reliable water supplies, and affordable access to markets if they are to maximize the economic and poverty-reducing benefits of low-cost drip systems.

Investing in water for livestock production

Livestock are an integral part of the socio-economic fabric of rural poor in all rural areas of SSA. They contribute to the livelihoods of the majority of the rural poor by strengthening their capacity to cope with income shocks (Ashley, Holden and Bazeley, 1999) and providing them with flexible access to cash when needed. Increasingly, global experience indicates that integrating water and livestock development creates more sustainable livelihoods zones and increases investment returns in ways that isolated development efforts are unlikely to produce (Molden, 2007).

Water-related investment to support livestock production varies from one livelihood zone to another as a function of the importance of livestock in the production system and of the prevailing climate conditions. In humid tropics, investment needs are limited as water sources are available for livestock, and livestock watering is not a particular concern. In more arid conditions, livestock watering issues become more relevant, while livestock play an increasingly important role in the livelihood zone. In relative terms, livestock are most important in arid, pastoral and agropastoral livelihood zones.

Easy access to an ample supply of water is a priority for livestock production. Regardless of how palatable and plentiful the forage or range may be, the livestock using it must have the water they need, or they will not thrive. Water deprivation quickly results in loss of appetite, and death occurs after a few days (3–5 days for zebus, 6–10 days for sheep, and 15 days or more for camels) when the animal has lost 25–30 percent of its weight (FAO, 1986). Inadequate stock water development in pastoral areas contributes to an unstable livestock industry and can lead to serious livestock losses. It also prevents profitable utilization of grazing areas and encourages destructive overgrazing in the vicinity of existing water supplies. In these systems, the development and maintenance of clean water supply systems for livestock is fundamental to enabling sustainable utilization of the forage without affecting the fragile equilibrium of the system.

There is a wide range of surface and groundwater water possibilities for stock water supply. Where conditions are ideal, one or more methods may be considered. The most likely locations for extending drinking-water from surface waters are where natural ponding already occurs. The cost of dug wells is usually high. However, involvement of the users in well digging has proved an efficient way to lower the cost of groundwater development. In many countries, stockbreeders tend to organize themselves through associations or cooperatives, which may be financially involved in groundwater development works (FAO, 1986).

Livestock water programmes need to be designed carefully. In the past, programmes that failed to take the livestock supporting capacity of rangeland adequately into account resulted in severe environmental damage and, in some cases, major problems of feed availability (FAO, 2006c), threatening the lives of entire herds.

Typically, the promotion of tubewell drilling in pastoral areas to enable the herds to stay longer in wet season grazing areas may lead to overgrazing, with long-term impacts on the ecology of the area.

Facilitating multiple use of water

In many areas, the volume of water available to households is as important as its quality. Households lacking sufficient water volume often do not implement sanitation practices that prevent the transmission of pathogens, such as washing hands and faces frequently (van der Hoek, Konradsen and Jehangir, 1999; Boelee, Laamrani and van der Hoek, 2007). However, improvements in water supply alone are unlikely to have positive health impacts unless sanitation practices are also improved. Optimal intervention programmes include improvements in water volume, water quality, and sanitation practices. However, the current understanding of water demand for productive uses is weak. Little is known about water use and demand in rural communities, and most of the research and development has focused on water for human consumption. Typically, water supply systems have been designed to provide small quantities of drinking-quality water at a relatively high price (Pérez de Mendiguren Castresana, 2003).

When possible, investments that provide water for more than one household purpose are likely to be more effective than single-purpose investments in improving livelihoods (Box 9). For example, constructing a village pond or investing in a community tubewell might provide water for irrigation, livestock production, and household chores. Such investments might also reduce the time required by household members to obtain water for drinking and other purposes from distant sources. Providing water of suitable quality nearer to homes and villages can reduce drudgery and enable household members to spend more time on productive activities. In Zimbabwe, many household wells provide sufficient water to support domestic uses and small-scale farming, which improves income and reduces poverty (Lane, 2004).

BOX 9
MULTIPLE USE OF DOMESTIC WATER IN SOUTH AFRICA

One study found a wide range of water-dependent productive activities in 13 communities in Bushbuckridge District, South Africa (Pérez de Mendiguren Castresana, 2003). Some of these activities provided goods and services to poor households, and they constituted an important element of the livelihoods of families. The main ones were: vegetable gardens, fruit trees, beer-brewing, brickmaking, hairdressing, livestock (cattle and goats), and ice-block making. Others included: grass-mat weaving; smearing and plastering of walls and floors; baking; poultry; and duck ponds.

Access to sufficient water is also essential for small agroprocessing, thus enhancing the value of agricultural production. This ranges from the simple washing of agricultural products to drying, packaging and canning. Health requirements for packed vegetables for export may also result in overall hygiene gains for the rural poor involved in such steps. Washing hands with soap leads to a significant reduction in intestinal diseases in families, and the packaged vegetables are not rejected by health inspectors.

In large irrigation schemes, people use water available in irrigation canals for multiple purposes. Canal water is often preferred to water from other sources for several reasons, including the volume available, accessibility, and practical considerations. Boelee, Laamrani and van der Hoek (2007) have identified five categories of water uses that are observed in irrigated areas other than irrigation of the main crops:

- agriculture-related purposes, such as irrigating home gardens, watering livestock, washing agricultural equipment, and soaking fodder;
- domestic purposes, such as laundry, bathing, washing household utensils, soaking grains, cooking, drinking, house cleaning, and sanitation;
- commercial purposes, usually small-scale activities or home industries, such as brickmaking, butcher's or other shops, washing vehicles, pottery, and mat weaving;

o productive purposes, usually non-consumptive, such as fisheries and water mills;

o recreation.

The additional benefits made possible by providing water for household purposes can enhance the aggregate value of investments in irrigation. In some areas, the additional benefits might produce a positive benefit–cost ratio for a project that might otherwise not generate a positive return.

Households and small commercial firms in SSA might also benefit from the development of aquaculture in conjunction with existing or new irrigation systems. The concept of integrated irrigation aquaculture (IIA) is extensively documented for West Africa and other regions where fish is produced in irrigation reservoirs and canals, or in irrigated rice fields (FAO, 2006d). Fish production and harvesting have been conducted both formally and informally in irrigation systems, in flood recession schemes, swamps, bas-fonds and small ponds in Africa and elsewhere for many years, providing an additional source of food and revenue for many households. Further development of aquaculture production, particularly in extensive small-scale irrigation settings, might enhance rural livelihoods and reduce household vulnerability while also improving the aggregate productivity of water resources. A number of commonly available agricultural by-products represent a potential source of feed, and the protein efficiency of fish is usually higher than that of other animals (Molden, 2007). In addition, sediments from small aquaculture ponds can be used as fertilizer in agriculture.

The main challenge, other than the production-related ones, concerns the customary and/or formal governance of the water bodies. Different users, with different power positions, use freshwater resources for different purposes at different times of the year, and throughout the years – sometimes with large intermittent periods of absence. Such multiple-use/multiple-user scenarios are under even more stress and more vulnerable to conflicts when droughts and floods place additional burdens on access to assets and distribution of benefits.

Addressing multiple needs for water has a strong gender aspect. Women and men often have different priorities for water use in a water management scheme.

While in most cases men use water for irrigating cash crops, women focus on growing staple/food crops and vegetables in home gardens, or use water for domestic purposes. The sustainability of a water management scheme for agricultural production may be at risk where other, sometimes conflicting, uses of water by women and men living in and close to the scheme are ignored (FAO, forthcoming).

If water management projects are to address concerns of both women and men, WUAs need to play an active role in localized water management for multiple use through recognizing the multiple uses of water in and around households for agriculture and for small-scale activities that allow both men and women to grow more crops and vegetables and to rear livestock.

ESSENTIAL CONDITIONS FOR SUCCESS

The likelihood of reducing poverty and improving food security in SSA through investments in the water sector depends on many supportive complementary investments in human, physical, financial, natural and social capital. The returns to major investments in new irrigation systems or investments that enhance rainfed production of staples or marketable crops will be small if farmers do not operate in a favourable environment. Markets, land tenure, property rights, water allocation procedures, and methods for resolving conflicts over land and water resources have substantial influence on the motivation, ability and success of smallholders in maximizing the value of investments in the water sector. Viable input and output markets, in which property rights are well defined and supported by the state, enable smallholders to obtain inputs and sell produce at competitive prices. Access to inputs and financial support, physical infrastructure, and investment in human capacities and technologies are also fundamental to the success of water development programmes. Discussed below are some of the key conditions for the success of water interventions in reducing poverty in rural SSA.

Ensuring enabling governance and policies

The policy environment must be supportive of smallholder production, consumption, and marketing of agricultural products. Policies at both macroeconomic and microeconomic level influence farm-level access to inputs and the ability to sell farm products at prices that provide sufficient revenue to sustain crop production. Macroeconomic policies must not create overvalued currency exchange rates that make exports more expensive, thereby reducing export opportunities for domestic farmers. Governments must also allow the importation of farm inputs and technological developments that might boost crop production at lower costs than is possible using only domestically produced inputs or existing production methods. Tariffs and quotas that restrict international trading of agricultural inputs and outputs must be considered carefully by public officials, as such limits can increase the cost of farming and reduce the revenues available to smallholders.

Policies regarding imports of food and fibre require particular attention. For many years, such imports, often arriving in the form of food aid from industrialized nations and international organizations, have increased the local supply in many countries of SSA. The increase in supply generally has had a downward impact on local prices, to the detriment of domestic farmers attempting to obtain market prices that cover their domestic costs of production. This impact discourages local farmers from investing in the quality or sustainability of soil and water resources, while also reducing labour opportunities in local economies.

The increases in urban populations that are occurring in many SSA countries and the global trends for rising agricultural food prices provide new opportunities for domestic farmers to increase production and receive attractive prices provided that the policy environment is supportive.

Policies that promote investments in local agricultural production will generate greater long-term benefits than efforts to increase imports of lower-cost food products available on international markets.

Governance has implications at all levels in agricultural water management. Table 3 shows the different governance dimensions corresponding to different scales of intervention and the need to address governance issues in relation to water, land, infrastructure and market services.

TABLE 3

Dimensions of governance and intervention

LEVEL	WATER	LAND	INFRASTRUCTURE	MARKET SERVICES
Farmer	Access to water: water rights; water markets	Access to land: land tenure; size of farm holdings	Access to affordable technology, including irrigation	Access to production inputs and markets
Farmer groups	Water rights; equity; water distribution; accountability	–	Management authority (irrigation schemes)	Farmer cooperatives, unions, meteorological forecasting
Irrigation service	Reliability, equity and flexibility of irrigation service delivery	Crop patterns and licensing	System management and maintenance; cost recovery; transparency; accountability	Farm roads maintenance and other scheme infrastructures
Local government	Water licensing (nepotism); conflict resolution	Land-use planning	Decentralization; development of new infrastructure (including markets)	Market infrastructure and transport; access to finance; market information
Basin authority	Sectoral water allocation; water quality management; water conservation (financial incentives)	Soil conservation; watershed protection	Main hydraulic infrastructure planning; development and management (corruption)	–
National government	Water policy and legislation; institutional arrangements	Land-use policy and legislation; cadastre; land-use planning	Policies and legislations on: decentralization; infrastructure development planning; cost recovery; financing mechanisms for infrastructure; access to finance for local stakeholders	Policies and legislation on: food security; agriculture (subsidies); rural development; trade (tariffs, subsidies); food self-sufficiency; rural finance
Regional	Transboundary water; security of supply	–	Transboundary water shared infrastructure	Regional trade agreements
Global level	International security and solidarity	–	–	Agricultural subsidies and tariffs

Source: WWAP (2006)

Securing access to markets

The effective operation of markets for food and agricultural products requires:

o appropriate legal frameworks and efficient institutions to support market conduct, the enforcement of contracts, and property rights;

o institutional frameworks for monitoring and supporting the emergence of markets through activities such as providing market information and marketing extension;

o well-operated and well-maintained infrastructure to provide transport and communication networks, post-harvest handling and storage, and physical markets.

Agricultural input and output markets must be accessible to smallholders, and information regarding input and output prices must be available to all participants. Smallholders can use new developments in communication technology to obtain current information describing input and output prices across a range of possible buyers and sellers. Public investments in regional communication networks can be helpful in providing smallholders with the access they need in order to optimize their participation in local and regional markets.

Many farmers in SSA have limited experience with formal, freely-functioning markets for agricultural inputs and outputs. Such a situation constrains public efforts to reduce poverty and improve food security through investments in the water sector. Hence, there is a role for government in training farmers to understand market operations and to help farmers produce and prepare their crops in ways that will enhance the likelihood of obtaining good prices in market settings. Extension service personnel can assist farmers in implementing measures that will improve the quality of farm products. Affordable access to farm chemicals, refrigeration, and transport services will also be helpful in this effort. Over time, public agencies might also assist farmers in forming cooperative associations that might provide additional services to members, such as promoting market development, exploring export opportunities, and seeking ways to add value to farm products before selling them in domestic or international markets. Farmers cooperatives could be based on, or form the basis for creating, effective WUAs. Water planners often consider forming WUAs when designing new irrigation schemes. Such associations could expand over time to undertake a variety of activities that support farm production and marketing. The goals of expansion might include providing additional services that enhance farm-level revenues, and generating additional funds to sustain the WUAs.

Physical infrastructure

Despite substantial investments in infrastructure in the recent past, rural populations in many countries of SSA remain poorly served. Inadequate investment in physical infrastructure limits the pace of economic development in many areas of SSA. Water supply, sanitation, and reliable electricity services are available in too few villages and districts. Paved roads, railroad networks, and easily accessible market centres are rare. In many countries, there are fewer than 1 000 km of paved roads per 1 000 persons, a level of service that is an order of magnitude smaller than the amount of paved roads in many industrialized nations.

Inadequate availability of storage, processing, refrigeration and packaging facilities are partly responsible for post-harvest losses that continue to be excessive in many rural areas (up to 30 percent of harvested fruit and vegetables), and limit opportunities for adding value to agricultural products. In situations where there is a food deficit, it is unacceptable to have post-harvest losses that can be avoided.

In many areas of SSA, investments in infrastructure will enhance the returns to investments in water control. The infrastructure needs are substantial, but so are the potential direct and indirect returns to appropriate investments. Infrastructure development is needed at all levels of investment:

o At the macrolevel, efforts should be made to ensure basic transport and communication infrastructure. Improved access and density of roads can reduce transaction costs for both inputs and outputs. Improvements in transportation, in particular when coupled with rural electrification, often lead to an increase in the cultivation of improved varieties of plants, increased fertilizer use, and expansion of areas under irrigation and water management. Transport and telecommunication services enable communication and information flow between rural and urban centres. This links farmers to markets and also facilitates the flow of information to and from extension specialists. The secondary and synergistic impacts of investments in roads, electricity and other forms of communication can be substantial, particularly in the least developed areas. The introduction of mobile phones has considerably increased information on markets for previously remote farmers and, thereby, increased their market opportunities. This changes the attractiveness of investment in various types of infrastructure.

o At the mesolevel, the development of safe and well-organized physical markets, both wholesale and retail, is important for facilitating the exchange of goods at regional level. In rural areas, markets not only provide a convenient location for farmers to meet with traders and consumers, they are also focal points for community activities. Some attempts to improve market infrastructure have been disappointing in the past, partly because of inadequate consultation with users. Better consultation might increase the likelihood of designing market centres that serve many purposes in ways that truly promote commerce and enhance the timely dissemination of market information.

o At the microlevel, investments in post-harvest handling, storage and processing facilities can also stimulate the non-farm sector and support the creation of small businesses. This can be a significant source of employment and, hence, income for poor people in rural areas.

The complementary nature of investments in irrigation and other forms of infrastructure, such as roads, schools, and health care facilities, is somewhat symmetric. As investments in roads and schools can improve the returns to investments in irrigation, so too can investments in irrigation improve the returns to investments in roads and schools (Ali and Pernia, 2003). It is reasonable to expect that the value of improving roads in a rural area will be greater if farmers have access to irrigation.

Land tenure and water rights

Farm-level efforts to improve and maintain productivity will be of limited value unless land tenure is secure for smallholders. Farmers must be able to count on the long-term benefits of near-term investments that reduce the rate of land degradation and maintain growth in productivity. In many areas of SSA, systems of land tenure and water-use rights have become dysfunctional and limit investment. Both land tenure and water rights issues must be addressed in a coordinated fashion in order to ensure optimal returns to public investments in irrigation and to motivate adequate investments at farm level.

Conflicts involving land and water resources often increase with population density and with increases in economic activity. In densely populated areas, the withdrawal of water for irrigation or other uses from the upper reaches of a river basin or watershed competes with the needs of people downstream. Effective river basin institutions are needed in such areas. Economic incentives might also be needed to achieve a socially optimal re-allocation of water, in conjunction with defining water rights to shifts in water allocation.

More generally, the environmental sustainability of rural investment is inextricably linked to the economic and social development of the recipient communities. Genuine ownership on the part of communities is the most effective path to environmental sustainability. Without these, the overall economic, social and environmental sustainability of the water infrastructure investment is at risk.

Preventing soil degradation and restoring fertility

Investments in the water sector will not be successful unless smallholders have affordable access to complementary inputs, in particular fertilizers (Box 10). The average annual rate of growth in fertilizer use in SSA declined from almost 9 percent between 1962 and 1982 to less than 1 percent between 1982 and 2002, partly because of the removal of fertilizer subsidies in the 1980s and 1990s.

BOX 10
THE ROLE OF FERTILIZERS IN CONTRACT FARMING

Farmers in some areas of SSA have opportunities to produce cash crops that are purchased by trading firms in accordance with contracts that describe production goals and crop prices. Contract farming arrangements often provide financial credit to farmers at the start of a production season. Participating farmers can intensify crop production by applying more fertilizer and other inputs than would be possible without credit. In some cases, the credit enables farmers to increase their use of fertilizer on both their cash crops and their food crops. Jayne, Yamano and Nyoro (2004) observed this result in a panel survey involving crop production data for 1 540 households in Kenya in the period 1997–2000. Households engaged in marketing arrangements for selected cash crops applied substantially more fertilizer on those crops and on cereal crops than did households not engaged in marketing arrangements.

Government involvement in the provision of seed, fertilizer, and chemicals lost favour with international organizations in the 1980s and 1990s. Structural adjustment programmes required governments to discontinue subsidizing farm inputs. As a result, average productivity declined. Estimated soil nutrient losses exceeded 60 kg/ha in 21 countries in SSA in 2002-04 (Table 4). Declining soil productivity reduces crop yields and sets in motion a vicious cycle that might be described as inadequate soil fertility causing low crop yields, which produce limited farm revenue, such that farmers lack funds for purchasing mineral fertilizers. As this cycle is repeated over time, soil fertility and crop yields continue to decline. Input subsidies are needed in some areas in order to restore growth in agricultural productivity and ensure the success of new interventions in the water sector.

Recently, governments that have restored an element of targeted fertilizer subsidy for the poor have seen gains in output and incomes in this group. This is discussed further below.

TABLE 4

Estimated soil nutrient losses in African countries, cropping seasons 2002–04

LOW (less than 30 kg/ha/year) (kg/ha)		MEDIUM (from 30 to 60 kg/ha/year) (kg/ha)		HIGH (more than 60 kg/ha/year) (kg/ha)	
Egypt	9	Libyan Arab Jamahiriya	33	United Republic of Tanzania	61
Mauritius	15	Swaziland	37	Mauritania	63
South Africa	23	Senegal	41	Congo	64
Zambia	25	Tunisia	42	Guinea	64
Morocco	27	Burkina Faso	43	Lesotho	65
Algeria	28	Benin	44	Madagascar	65
		Cameroon	44	Liberia	66
		Sierra Leone	46	Uganda	66
		Botwana	47	Democratic Republic of the Congo	68
		Sudan	47	Kenya	68
		Togo	47	Central African Republic	69
		Côte d'Ivoire	48	Gabon	69
		Ethiopia	49	Angola	70
		Mali	49	Gambia	71
		Djibuti	50	Malawi	72
		Mozambique	51	Guinea Bissau	73
		Zimbabwe	53	Namibia	73
		Niger	56	Burundi	77
		Chad	57	Rwanda	77
		Nigeria	57	Equatorial Guinea	83
		Eritrea	58	Somalia	88
		Ghana	58		

Source: Henao and Baanante (2006)

Providing targeted subsidies and adapted financial packages

Focusing on agriculture, the World Development Report 2008 (World Bank, 2007b) acknowledges the importance of well-targeted input subsidies as an element of poverty reduction strategies in rural areas. Several mechanisms are available to support farm-level purchases of key inputs, from providing selected inputs at no charge to farmers to low-interest-bearing seasonal or mid-term loans. The optimal combination of available methods will vary among countries and among production regions.

The goal in all cases should be to ensure affordable access to infrastructure, services and inputs, particularly for smallholders who are most vulnerable to shortfalls in agricultural production. Public assistance for purchasing key inputs will impose a cost on governments, while lowering the farm-level cost of producing crops and livestock products. The public cost can be justified by the non-market, public benefits of boosting agricultural production in a comprehensive effort to reduce poverty and improve food security (Box 11).

In addition to credit for purchasing the inputs needed at the start of each crop season, farmers must also have access to the financial credit needed to make investments that will generate benefits over time. Developers of financial tools and packages to support water investments in rural areas need to recognize the many different functions of water for agriculture and the spectrum of possible water interventions. The variety of functions and the range of possible interventions provide scope for designing innovative programmes that correspond to specific needs. For example, term finance needs to be promoted to support medium-term water-related investments. Figure 4 shows how different social groups require specific financial support.

BOX 11
RECORD MAIZE HARVEST IN MALAWI

Malawi has a chronic hunger problem, with more than one-fifth of the population unable to meet their daily food needs. One cause of the food shortage has been the poor crop harvests that the country has suffered for many years. In the last two years (2006 and 2007), the country has experienced bumper harvests, with a surplus of 1 million tonnes of maize in 2007. Behind these record results is the Government of Malawi's fertilizer and seed subsidy programme, introduced in 2005 and cofunded by the Department for International Development (DFID) of the United Kingdom. This programme, which allows Malawians to buy fertilizer and maize seed at better prices than in the past, has benefited some of the country's poorest people. In the future, the programme should help secure Malawi's food supplies in a sustainable way, while providing smallholder farmers with improved sources of livelihood.

Source: DFID (2007)

FIGURE 4

Adapting financial services to the needs of different groups

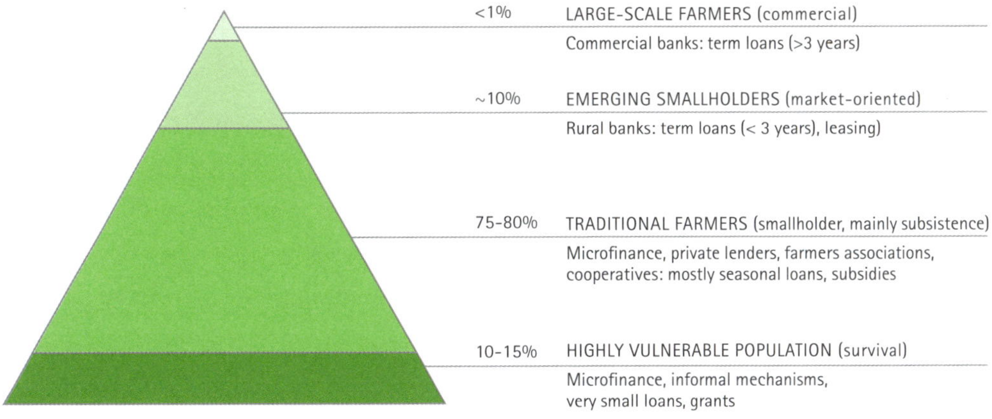

<1%	LARGE-SCALE FARMERS (commercial)
	Commercial banks: term loans (>3 years)
~10%	EMERGING SMALLHOLDERS (market-oriented)
	Rural banks: term loans (< 3 years), leasing
75-80%	TRADITIONAL FARMERS (smallholder, mainly subsistence)
	Microfinance, private lenders, farmers associations, cooperatives: mostly seasonal loans, subsidies
10-15%	HIGHLY VULNERABLE POPULATION (survival)
	Microfinance, informal mechanisms, very small loans, grants

Investing in human capital

Complementary investments in education and training enhance the value of investments in irrigation and water control by providing farmers with appropriate knowledge and skills. Similarly, the returns to investments in education and training will be higher if farmers have opportunities to implement new production methods in irrigated settings.

Within this context, it is necessary to consider the important roles of women in irrigation, water harvesting, and other aspects of agricultural production in developing countries. The concerns of women must be taken into account in the conceptual phase of water investment projects. Excluding women from the design phase may have unexpected adverse effects in terms of poverty reduction and equity (FAO, forthcoming). For example, an inappropriate design or location of tap-stands or wells may lead inadvertently to an increase in burdens or safety concerns for women and young girls charged with fetching the water. Similarly, a tight water rotation schedule is usually not suitable for women who must perform many different domestic tasks and do not have full control over their time. Therefore, capacity-building programmes in water management should be designed in ways that relieve women and girls from part of the heavy burden in conducting daily tasks.

ADAPTING INTERVENTIONS TO LOCAL CONDITIONS

Not all intervention options have the same relevance and potential for poverty reduction in all settings. As stated throughout this report, agroclimatic conditions, prevailing livelihood zone types, and local socio-economic conditions all influence intervention programmes. Table 5 provides a summary of the relevance of the main intervention options described above in different livelihood contexts. While it can be further refined to take into account local conditions, it shows that, at regional level, substantial differences in patterns of investments can be observed in different regions. Table 5 also confirms the results showing the potential for water intervention by livelihood zone, with particular emphasis on cereal-based and agropastoral zones.

TABLE 5

Relevance of intervention by livelihood zone

Livelihood zone	Manage soil moisture in rainfed areas	Invest in small-scale water harvesting infrastructure	Promote small-scale community-based irrigation	Improve existing irrigation systems	Improve water control for peri-urban producers	Invest in water for livestock production	Facilitate multiple use of water
Arid	low	moderate	low	High in irrigated schemes, n/a elsewhere	High around cities	low	low
Pastoral	low	low	low			high	high
Agropastoral	moderate	moderate	moderate			high	high
Cereal based	high	high	high			moderate	moderate
Cereal-root crop	moderate	moderate	high			moderate	moderate
Root-crop-based	low	low	moderate			low	moderate
Highland Temperate	high	moderate	moderate			moderate	moderate
Highland Perennial	low	moderate	moderate			moderate	moderate
Tree crop	low	low	low			low	moderate
Forest-based	low	low	low			low	moderate
Large Commercial and Smallholder	moderate	moderate	moderate			moderate	moderate
Rice-tree crop	low	low	moderate			low	moderate
Coastal Artisanal Fishing	low	low	low			low	moderate
Expected benefits (direct, by category of farmers)							
Large-scale	low	low	low	medium	low	low	low
Emerging	low	medium	medium	medium	high	medium	low
Traditional	high	high	high	low	low	high	high
Highly vulnerable	low	low	low/medium	low	low	medium	high

Source: Compilation of information based on FAO and International Fund for Agricultural Development (IFAD) projects

Soil moisture management, and in particular conservation agriculture practices, are most relevant in cereal–root crop zones and in highland temperate zones, where they can contribute to reducing the impact of dry spells in an otherwise favourable rainfall environment. Water harvesting, in particular for supplementary irrigation, is highly relevant in cereal-based zones, especially those dominated by maize. Small-scale community-based irrigation finds its application in several settings, in particular those where rainfall alone cannot guarantee agricultural production. Investment in water control for livestock production is of most importance in arid and semi-arid environments.

ASSESSING INVESTMENT POTENTIAL

This section presents the results of an exercise to estimate the possible costs of a programme of investments in water in support of rural livelihoods. It is based on an assessment of the potential application of each of the seven water intervention options described above.

In line with the philosophy of this report, the proposed investments are expected to affect the livelihoods of rural people through increased water security and improved access to water for both domestic and productive purposes, increased resilience to climate shocks, and a consequent reduction in people's vulnerability. Such improvements in rural people's livelihoods will come from improved control of water for their main source of food and revenues, from reduced hardship in terms of working conditions and a consequent increase in labour productivity, and from improved hygiene and health conditions.

To this effect, the benefits to be expected from such investments can hardly be expressed solely in terms of increased production. They also need to account for reduced variability in production, gender empowerment, enhanced labour productivity, reduced burden of diseases, improved institutional capacities, etc. For this reason, the cost estimates of potential investments presented here are not accompanied with estimates of benefits.

In order to ensure consistency with the approach proposed in this report, the assessment used the following three criteria: prevalence of poverty; water as a limiting factor for rural livelihoods; and potential for water intervention (Annex 2 provides details of the methodology). The assessment at regional level consisted of the following steps:

- Potential for water intervention: for each of the seven categories of interventions, and for each livelihood zone, assessment of the maximum possible extent of application of the intervention, taking into account the rural population, cultivated area, and available water resources, in modalities that vary from one type of intervention to another;

- Water as a limiting factor: application of a coefficient taking into account the importance of water as a limiting factor for each livelihood zone;

- Poverty incidence: application of a coefficient taking into account the importance and incidence of poverty for each livelihood zone.

- Unit costs by type of intervention were estimated based on available information from investment projects used by FAO for similar regional assessments. These unit cost figures represent only rough averages. Substantial differences can be expected from one livelihood zone to another, and from one place to another within a given zone.

The results are presented in detail in Annex 2 and in summary form in Tables 6–8. Table 6 shows the potential for each type of intervention by livelihood zone. It is expressed in potential area of rainfed and irrigated land, required storage capacity, heads of livestock and number of households reached, according to the type of intervention.

TABLE 6

Potential for water-related interventions by livelihood zone

Livelihood zone	Manage soil moisture in rainfed areas	Invest in small-scale water harvesting infrastructure	Promote small-scale community-based irrigation	Improve existing irrigation systems	Improve water control for peri-urban producers	Invest in water for livestock production	Facilitate multiple use of water
	ha	mm³	ha	ha	ha	head	household
Arid	114 770	34	30 000	389 793	62 606	1 255 260	250 272
Pastoral	8 948 023	2 684	500 000	601 019	113 497	24 223 700	4 904 028
Agropastoral	41 547 366	12 464	600 000	458 437	234 625	35 174 400	6 917 706
Cereal based	35 413 458	10 624	499 407	312 130	322 533	24 497 200	11 862 252
Cereal-root crop	51 176 547	15 353	358 122	223 826	249 844	38 576 100	12 229 596
Root-crop-based	2 146 486	644	11 192	93 267	111 223	1 218 008	730 676
Highland Temperate	7 576 418	2 273	104 128	86 774	123 970	9 283 125	4 054 523
Highland Perennial	1 756 652	527	10 772	26 930	80 667	1 563 705	1 637 755
Tree crop	305 265	92	2 087	57 965	94 816	94 189	133 312
Forest-based	818 626	246	5 491	45 758	73 991	249 578	437 555
Large Commercial and Smallholder	2 077 440	623	0	709 010	118 778	1 924 965	613 173
Rice-tree crop	150 575	45	6 501	346 763	15 261	86 510	120 785
Coastal Artisanal Fishing	73 299	22	6 724	186 787	103 205	44 258	70 011
Total	152 104 925	45 631	2 134 424	3 538 456	1 705 016	138 190 997	43 961 643

Table 7 estimates the number of rural people who can be reached in each livelihood zone by the type of intervention – the assessment considered persons rather than households (therefore, that what benefits a smallholder farmer benefits the whole family). The interventions are not all mutually exclusive. Thus, it can be expected that a person may benefit from one or more of the proposed investments. In total, it is expected that about 58 percent of the rural population of SSA could benefit from some type of investment in water. The percentage varies from 96 percent in the cereal-based area, to a few percentage points in areas where such interventions are not economically or socially justified.

TABLE 7

Number of people reached by intervention and livelihood zone

Livelihood zone	Manage soil moisture in rainfed areas	Invest in small-scale water harvesting infrastructure	Promote small-scale community-based irrigation	Improve existing irrigation systems	Improve water control for peri-urban producers	Invest in water for livestock production	Facilitate multiple use of water	Total	Total in % of rural population
	no. people	no. people	no. people	no. people	no. people	no. people	no. people	no. people	%
Arid	61 983	18 595	300 000	3 897 930	626 065	1 126 223	1 251 359	4 885 977	59
Pastoral	2 401 811	720 543	5 000 000	6 010 185	1 134 968	24 520 140	24 520 140	24 520 140	90
Agropastoral	18 800 948	5 640 284	6 000 000	4 584 370	2 346 248	30 745 360	34 588 530	34 588 530	90
Cereal-based	51 807 865	15 542 359	4 994 072	3 121 295	3 225 328	32 950 700	59 311 260	63 148 560	96
Cereal-root crop	53 882 439	16 164 732	3 581 216	2 238 260	2 498 440	33 971 100	61 147 980	62 200 355	92
Root-crop-based	2 903 776	871 133	111 920	932 665	1 112 228	1 461 351	3 653 378	5 060 589	10
Highland Temperate	17 715 750	5 314 725	1 041 282	867 735	1 239 704	9 010 050	20 272 613	20 864 471	69
Highland Perennial	6 501 187	1 950 356	107 720	269 300	806 670	3 275 510	8 188 775	8 188 775	25
Tree crop	528 729	158 619	20 867	579 645	948 156	266 623	666 558	2 077 397	7
Forest-based	1 518 707	455 612	54 909	457 575	739 912	875 109	2 187 773	2 771 103	9
Large Commercial and Smallholder	1 946 780	584 034	0	7 090 095	1 187 784	1 532 933	3 065 865	10 224 659	50
Rice-tree crop	359 095	107 729	65 015	3 467 630	152 606	241 571	603 926	4 044 346	50
Coastal Artisanal Fishing	157 022	47 107	67 243	1 867 870	1 032 052	140 023	350 057	3 124 188	20
Total	158 586093	47 575 828	21 344 244	35 384 555	17 050 161	140 116 692	219 808 213	245 699 091	58

Note: Total per livelihood zone is lower that the total of single interventions because some people will benefit from several types of intervention

TABLE 8

Investment costs by intervention and livelihood zone

Livelihood zone	Manage soil moisture in rainfed areas	Invest in small scale water harvesting infrastructure	Promote small-scale community-based irrigation	Improve existing irrigation systems	Improve water control for peri-urban producers	Invest in water for livestock production	Facilitate multiple use of water	Total	Total per beneficiary	Total per ha of cultivated land (*)
	USD million	USD million	USD million	USD million	USD million	USD million	USD million	USD million	USD/pers.	USD/ha
Arid	9	34	128	780	188	38	19	1 194	244	737
Pastoral	671	2 684	2 125	1 202	340	727	368	8 118	331	692
Agropastoral	3 116	12 464	2 550	917	704	1 055	519	21 325	617	465
Cereal-based	2 656	10 624	2 122	624	968	735	890	18 619	295	472
Cereal-root crop	3 838	15 353	1 522	448	750	1 157	917	23 985	386	424
Root-crop-based	161	644	48	187	334	37	55	1 464	289	48
Highland Temperate	568	2 273	443	174	372	278	304	4 412	211	373
Highland Perennial	132	527	46	54	242	47	123	1 170	143	141
Tree crop	23	92	9	116	284	3	10	537	258	38
Forest-based	61	246	23	92	222	7	33	684	247	58
Large Commercial and Smallholder	156	623	0	1 418	356	58	46	2 657	260	167
Rice-tree crop	11	45	28	694	46	3	9	835	206	305
Coastal Artisanal Fishing	5	22	29	374	310	1	5	746	239	204
Total	11 408	45 631	9 071	7 077	5 115	4 146	3 297	85 745	349	334
Total in percentage of total cost	13	53	11	8	6	5	4	100		

Note: (*): The total per hectare of cultivated land refers to the first five interventions

Table 8 expresses these potential interventions in terms of capital investment costs. In total, these investments could amount to about USD 86 000 million, which would represent USD 350 per beneficiary. For land-related interventions, the average investment would be about USD 330/ha. The bulk of the costs (53 percent) would be for small-scale water harvesting infrastructures, in support of supplementary irrigation and other uses such as fish farming. This category of intervention is broad and ranges from very small check dams to small reservoirs and subsurface reservoirs. Soil moisture management in rainfed areas and small-scale community-based irrigation also represent substantial potential. Of lower value in terms of investment costs, but locally important, are interventions such as livestock watering and the development of multiple-use systems.

These figures should be taken as being only indicative and as an order of magnitude of the potential for investments in water in support of rural poverty reduction in SSA. Considerable uncertainties are associated with the estimation of "average" unit costs, and of the extent of the potential of each intervention. In particular, the range of options captured under the heading "small-scale water harvesting" and the range of costs associated with these interventions, together with the extent of possible application of such investments, are the single most important factor influencing the estimates of costs.

CONCLUSION

This report carries two important messages. The first is that there is a large range of opportunities for interventions in water in support of the rural poor in SSA. The potential for such interventions in terms of people reached, water mobilized and land productivity enhancement is extremely large. In total, it is estimated that about 58 percent of the rural population of SSA could benefit from some type of investment in water. Water will remain a major factor affecting the livelihoods of rural people in the region, both in terms of basic services, and in terms of resilience building and vulnerability reduction. However, as advocated here, these water interventions are unlikely to generate poverty reduction effects if they are conducted in isolation, without also acting on the political, institutional, market, knowledge, and financial dimensions of the challenge.

The second message is that the variety of livelihood situations in which rural people operate in SSA calls for context-specific and targeted interventions, where rural people's constraints and opportunities are understood and addressed, and where they can take part in the decision-making processes in a way that is effective and ensures the greatest impact on their livelihoods. While all categories of rural people are expected to benefit directly or indirectly from such interventions, the traditional smallholders, farmers, fishers and herders offer the greatest potential for poverty reduction.

Rural communities are in transition, and the dynamics of this transition need to be understood and internalized in order to design effective poverty reduction programmes. As a basic human need, and as a major production factor in rural areas, water has a central role to play in helping rural communities to meet new challenges and to benefit from the associated opportunities.

REFERENCES

Abalu, G. & Hassan, R. 1998. Agricultural productivity and natural resource use in southern Africa. *Food Pol.*, 23(6): 477–490.

Ali, I. & Pernia, E.M. 2003. *Infrastructure and poverty reduction: what is the connection?* Economics and Research Department Policy Brief Series, No. 13. Manila, ADB.

Ashley, S.D., Holden, S.J & Bazeley, P.B.S. 1999. *Livestock in development 1999. Livestock in poverty focused development.* Crewkerne, UK.

Aw, D. & Diemer, G. 2005. *Making a large irrigation scheme work: a case study from Mali.* Washington, DC, World Bank.

Baïdu-Forson, J. 1999. Factors influencing adoption of land-enhancing technology in the Sahel: lessons from a case study in Niger. *Agric. Econ.*, 20(3): 231–239.

Boelee, E., Laamrani, H. & van der Hoek, W. 2007. Multiple use of irrigation water for improved health in dry regions of Africa and South Asia. *Irri. Drain.*, 56(1): 43–51.

Dalton, T.J. & Guei, R.G. 2003. Productivity gains from rice genetic enhancements in West Africa: countries and ecologies. *World Dev.*, 31(2): 359–374.

Department for International Development (DFID). 2007. *Record maize harvest in Malawi. Case studies* (available at http://www2.dfid.gov.uk)

Diao, X. & Pratt, A.N. 2007. Growth options and poverty reduction in Ethiopia: an economy-wide analysis. *Food Pol.*, 32(2): 205–228.

Diemer, G. & Huibers, F.P., eds. 1996. Crops, people and irrigation: water allocation practices of farmers and engineers. London, IT Publications.

Drechsel, P. & Varma, S. 2007. *Recognizing informal irrigation in urban and peri-urban West Africa.* Water Policy Briefing, Issue 26. Colombo, IWMI. 8 pp.

Drechsel, P., Graefe, S., Sonou, M. & Cofie, O.O. 2006. *Informal irrigation in urban West Africa: an overview.* Research Report 102. Colombo, IWMI. 34 pp.

Erenstein, O., Oswald, A. & Mahaman, M. 2006. Determinants of lowland use close to urban markets along an agro-ecological gradient in West Africa. *Agric. Ecosys. Env.*, 117(2–3): 205–217.

FAO. 1986. *Water for animals*, by P. Pallas. FAO Land and Water Miscellaneous. AGL/ MISC/4/85. Rome.

FAO. 1994. *Water harvesting for improved agricultural production.* Proc. FAO Expert Consultation, 21–25 November 1993, Cairo. Water Report No. 3. Rome.

FAO. 1998. *Guidelines for the water control component of the Special Programme for Food Security.* Unpublished document.

FAO. 2001. *The economics of conservation agriculture.* FAO Land and Water Development Division. Rome.

FAO. 2002. Project GCP/NER/032/ITA. Project integer Keita. Rapport terminal du projet. Rome.

FAO. 2005. *Drought-resistant soils. Optimization of soil moisture for sustainable soil production.* FAO Land and Water Bulletin No. 11. Rome.

FAO. 2006a. *Demand for irrigated products in sub-Saharan Africa.* Water Report No. 31. Rome.

FAO. 2006b. *Irrigation in Africa in figures.* Rome.

FAO. 2006c. *Access to water, pastoral resource management and pastoralists' livelihoods*, by N. Gomes. FAO Livelihood Support Programme (LSP). Rome.

FAO. 2006d. *Integrated irrigation and aquaculture in West Africa: concepts, practices, and potential*, by M. Halwart & A.A. van Dam, A.A., eds. Rome.

FAO. forthcoming. *Pocket guide for gender mainstreaming in water management.*

FAO-AQUASTAT. 2008. Online database on water in agriculture (available at http://www.fao.org).

FAOSTAT. 2008. FAO online database (available at http://faostat.fao.org).

Fox, P. & Rockström, J. 2000. Water harvesting for supplemental irrigation of cereal crops to overcome intra-seasonal dry-spells in the Sahel. *Phys. Chem. Earth*, (B) 25(3): 289–296.

Godswill, M., Kelemework, D. & Aredo, D. 2007. A comparative analysis of rainfed and irrigated agricultural production in Ethiopia. *J. Irri. Drain. Sys.* 21(1): 35–44.

Haefele, S.M., Johnson, D.E., Diallo, S., Wopereis, M.C.S. & Janin, I. 2000. Improved soil fertility and weed management is profitable for irrigated rice farmers in Sahelian West Africa. *Field Crops Res.* 66(2): 101–113.

Haefele, S.M., Wopereis, M.C.S., Donovan, C. & Maubuisson, J. 2001. Improving the productivity and profitability of irrigated rice production in Mauritania. *Eur. J. Agron.*, 14(3): 181–196.

Henao, J. & Baanante, C. 2006. Agricultural production and soil nutrient mining in Africa: implications for resource conservation and policy development. Muscle Shoals, USA, IFDC.

International Fund for Agricultural Development (IFAD). 2005. *Agricultural water development for poverty reduction in Eastern and Southern Africa.* Draft report on component study for the AfDB, FAO, IFAD, IWMI, and the World Bank Collaborative Programme Investment in Agricultural Water for Poverty Reduction and Economic Growth in Sub-Saharan Africa. Rome.

Inocencio, A., Kikuchi, M., Tonosaki, M. Maruyama, A., Merrey, D., Sally, H. & de Jong, I. 2007. *Costs and performance of irrigation projects: a comparison of sub-Saharan Africa and other developing regions.* IWMI Research Report 109. Colombo, IWMI. 71 pp.

Jayne, T.S., Yamano, T. & Nyoro, J. 2004. Interlinked credit and farm intensification: evidence from Kenya. *Agric. Econ.*, 31: 209–218.

Keller, J. & Roberts, M. 2004. Household-level irrigation for efficient water use and poverty alleviation. *In* V. Seng, E. Craswell, S. Fukai & K. Fisher, eds. *Water in agriculture*. ACIAR Proceedings No. 116. Canberra, ACIAR.

Kijima, Y., Sserunkuuma, D. & Otsuka, K. 2006. How revolutionary is the "NERICA Revolution?" Evidence from Uganda. *Dev. Econ.*, 44(2): 252–267.

Kulecho, I.K. & Weatherhead, K. 2006. Issues of irrigation of horticultural crops by smallholder farmers in Kenya. *Irri. Drain. Sys.*, 20: 259–266.

Lane, J. 2004. Positive experiences from Africa in water, sanitation and hygiene. *Water Pol.*, 6(2): 153–158.

Magistro, J., Roberts, M., Haggblade, S., Kramer, F., Polak, P., Weight, E. & Yoder, R. 2007. A model for pro-poor wealth creation through small-plot irrigation and market linkages. *Irri. Drain.*, 56(2–3): 321–334.

Molden, D. ed. 2007. *Water for food, water for life. A comprehensive assessment of water management in agriculture.* London, Earthscan, and Colombo, IWMI.

Morardet, S., Merrey, D.J., Seshoka, J. & Sally, H. 2005. *Improving irrigation project planning and implementation processes in Sub-Saharan Africa: diagnosis and recommendations.* Working Paper 99. Colombo, IWMI. 91 pp.

Mupangwa, W., Love, D. & Twomlow, S. 2006. Soil-water conservation and rainwater harvesting strategies in the semi-arid Mzingwane Catchment, Limpopo Basin, Zimbabwe. *Phys. Chem. Earth*, Parts A/B/C 31(15–16): 893–900.

Otsuka, K. & Kalirajan, K.P. 2006. Rice green revolution in Asia and its transferability to Africa: an introduction. *Dev. Econ.*, 44(2): 107–122.

Oweis, T., Prinz, D. & Hachum, A. 2001. *Water harvesting: indigenous knowledge for the future of the drier environments.* Aleppo, Syrian Arab Republic, International Center for Agricultural Research in the Dry Areas.

Pérez de Mendiguren Castresana, J. C. 2003. *Productive uses of water at the household level: evidence from Bushbuckridge, South Africa.* International Symposium on Water, Poverty and Productive uses of Water at the Household Level, 21–23 January 2003, Muldersdrift, South Africa. (available at http://www.wca-infonet.org).

Poussin, J.C., Wopereis, M.C.S., Debouzie, D. & Maeght, J.L. 2003. Determinants of irrigated rice yield in the Senegal River valley. *Eur. J. Agron.*, 19(2): 341–356.

Renwick, M. 2007. *Multiple use water services for the poor: assessing the state of knowledge. Winrock* international: Arlington, VA.

Roberts, N. 1988. Dambos in development: management of a fragile ecological resource. *J. Biogeog.*, 15(1): 141–148.

Rockström, J. 2000. Water resources management in smallholder farms in Eastern and Southern Africa: an overview. *Phys. Chem. Earth*, (B) 25(3): 275–283.

Rockström, J., Hatibu, N., Oweis, T. & Wani, S. 2007. Managing water in rainfed agriculture. *In* D. Molden, D. ed. 2007. *Water for food, water for life. A comprehensive assessment of water management in agriculture.* London, Earthscan, and Colombo, IWMI.

Rossi, B. 2005. *Development brokers and translators: the ethnography of aid.* London, School of Oriental and African Studies, University of London.

Sakurai, T. 2006. Intensification of rainfed lowland rice production in West Africa: present status and potential green revolution. *Dev Econ.*, 44(2): 232–251.

Van der Hoek, W., Konradsen, F. & Jehangir, W.A. 1999. Domestic use of irrigation water: health hazard or opportunity? *Int. J. Wat. Res. Dev.*, 15(1–2): 107–119.

Wopereis-Pura, M.M., Watanabe, H., Moreira, J. & Wopereis, M.C.S. 2002. Effect of late nitrogen application on rice yield, grain quality and profitability in the Senegal River valley. *Eur. J. Agron.*, 17(3): 191–198.

World Bank. 2007a. *Investment in agricultural water for poverty reduction and economic growth in sub-Saharan Africa.* Synthesis Report. A collaborative programme of AfDB, FAO, IFAD, IWMI and the World Bank.

World Bank. 2007b. World development report 2008. Agriculture for development. Washington, DC, World Bank.

World Water Assessment Programme (WWAP). 2006. *Water, a shared responsibility.* The United Nations World Water Development Report 2. World Water Assessment Programme. Paris, UNESCO, and New York, USA, Berghahn Books.

ANNEX 1
DESCRIPTION OF THE LIVELIHOOD ZONES USED

This annex provides a description of the prevailing conditions and main farming activities that sustain rural livelihoods in 13 main zones, plus two locally relevant livelihood zones. In the text below, the term "region" refers to sub-Saharan Africa (SSA).

Arid zone

This zone is the largest (21 percent of the region) and corresponds to the deserts of the Sahara and southwestern Africa. It has marginal importance in terms of agriculture and population. The area under cultivation covers only 0.3 percent of the land area of the livelihood zones (mostly oases), while the rural population (8 million) represents only 2 percent of the regional total. In view of the high level of aridity, irrigated areas represent almost half the cultivated land. Rangeland and livestock are confined to marginal areas. Living conditions are extremely hard, and the rural population consists mainly of nomads, and a few sedentary people at the oases.

Pastoral zone

This zone is located mostly in the semi-arid zones extending across the Sahel from Mauritania to the northern parts of Mali, Niger, Chad, Sudan, Ethiopia and Eritrea. Some parts are also found in northern Kenya and Uganda, and in part of Namibia, Botswana and southern Angola. It occupies almost 2.7 million km^2, or 11 percent of the area of the region. The rural population is 27 million (7 percent), with 24 million head of livestock. Pastoral land is abundant (more than 190 million ha). This zone is characterized by nomadic pastoralists, who move to other zones during the driest period of the year, and exclusive pastoralists. The latter are livestock producers who grow no crops and simply depend on the sale or exchange of animals and their products to obtain foodstuffs. Such producers are most likely to be nomads, i.e. their movements are opportunistic and follow pasture resources in a pattern that varies from year to year. This type of nomadism reflects, almost directly, the availability of forage resources – the patchier these are, the more likely an individual herder is to move in an irregular pattern.

Pastoralists are highly vulnerable to climate variability and droughts. In particular, they are highly dependent on the availability of water points for their animals. Fragile balances exist between the availability of water and feed for animals. In periods of drought, excessive concentration of animals around watering points may lead to catastrophic losses of herds. Some of Africa's largest irrigated areas are located in the pastoral zones of the Nile and Niger Rivers, such as Gezira Scheme in the Sudan, where integration of irrigated agriculture and livestock play an important role in overall agricultural production.

Agropastoral zone

This zone covers 2.15 million km², or 9 percent of the land of the region. It is characterized by a semi-arid climate, with an average growing period of 95–100 days. It extends from Senegal to Niger in West Africa, and covers substantial areas of East and Southern Africa from Somalia and Ethiopia to South Africa. The rural population represents 9 percent of the region accounting for more than 38 million people, with a density of 18 inhabitants/km². Although the population density is limited, pressure on fragile land is high. Field crops and livestock are equally relevant in the household livelihoods of this zone. Cultivated land and livestock account for 40 million ha and 35 million head, respectively, i.e. 18 and 19 percent of the regional total. Pastoral areas are abundant (more than 148 million ha) and represent 14 percent of the regional total and 70 percent of the area of the zone. Rainfed sorghum and millet are the main sources of food, which are rarely sold on local markets, while sesame and pulses are sometimes marketed. Cultivation is frequent along riverbanks, particularly alongside the Niger and Nile Rivers. Livestock is used for subsistence, marketing (milk and milk products), offspring, transportation, land preparation, sale or exchange, savings, bridewealth, and insurance against crop failure. The region is characterized by extremely low soil fertility and chronic limitations in terms of organic matter.

Irrigation plays a relatively important role in this zone, with more than 900 000 ha of recorded irrigated areas, putting substantial pressure on the region's water resources (20 percent of total water resources of the zone are diverted for irrigation). Rainfed cultivation is often accompanied by water conservation practices in an attempt to enhance soil moisture retention (zai, half-moons, stone ridges, etc.). Nonetheless, vulnerability to drought remains high, with frequent crop failures and deprived livestock.

Cereal-based zone

This livelihood zone covers large parts of the region (2.45 million km^2) and it is the most important food production zone in East and Southern Africa. It extents mainly along the Rift Valley, across plateau and highland areas at altitudes of 800–1 500 m, from Kenya and the United Republic of Tanzania to Zambia, Malawi, Zimbabwe, South Africa, Swaziland and Lesotho. The climate ranges from dry subhumid to moist subhumid. The cultivated area covers 36 million ha and accounts for 15 percent of the regional total. The rural population is almost 66 million, 16 percent of the regional total. Most of the zone has monomodal rainfall, but some areas experience bimodal rainfall. Farmers are typically traditional or emerging smallholders, with farms of less than 2 ha. The main crops are maize (staple and cash crop), tobacco, coffee and cotton. Yields have fallen in recent decades owing to the shortages and high cost of inputs such as seeds, fertilizers and agrochemicals. Soil fertility has been declining, prompting smallholders to revert to more to extensive production practices. About 24.5 million ruminants are kept both for food and farm manure and ploughing, and savings. In spite of scattered settlement patterns, community institutions and market linkages in the maize belt are relatively more developed than in other livelihood zones.

Small-scale irrigation schemes and supplementary irrigation are scattered within the zone, and cover 620 000 ha, or 9 percent of the regional total, although the potential is much higher. In this zone, a combination of soil fertility restoration and supplementary irrigation has the potential to boost agricultural productivity substantially, in response to rapidly decreasing farm size.

Cereal–root crop zone

This livelihood zone extends from Guinea through northern Côte d'Ivoire to Ghana, Togo, Benin and the mid-belt states of Nigeria to northern Cameroon, and on to Central and Southern Africa. It covers 3.17 million km^2 (13 percent of the land area of the region) – mainly in the moist semi-arid zone with an average growing period of about 130 days. Some 51 million ha (22 percent of the regional total) are cultivated, sustaining a rural population of almost 68 million (16 percent of the regional total). Livestock (mostly ruminants) are abundant (42 million head). Pasture, with almost 195 million ha, accounts for 18 percent of the regional area. Compared with the cereal-based zone, this zone is characterized by lower altitude, higher temperatures, lower population density, abundant cultivated land, and higher livestock numbers per household. It also has poorer transport and communications infrastructure. Cereals such as maize, sorghum and millet are common in the area, rotated or intercropped with root crops such as yams, cassava and sweet potatoes. Although a range of agricultural products are marketed, most of the products are consumed within households, given the prevalence of subsistence agriculture and traditional farmers.

Irrigation is limited, it accounts for 6 percent of the regional total, with fewer than 422 000 ha, despite a relatively high potential, estimated at 7.7 million ha. A range of water intervention options have potential for poverty reduction, in particular soil moisture management practices, supplementary irrigation and community-level small-scale irrigation.

Root-crop-based zone

This livelihood zone corresponds mainly to a subhumid climate. It covers 2.8 million km² (about 11 percent of the land area of the region), has a cultivated area of 28 million ha, and is home to 48 million rural people. Precipitation patterns show a good seasonal distribution, and the risks of crop failure are limited. The zone contains about 16 million head of livestock. Farmers are mainly traditional smallholders, typically oriented towards staple crops and self-consumption, and root crops are indeed the main staple. Market prospects exist in places, in particular for export of oil-palm products, urban demand for root crops is growing, and linkages between agriculture and off-farm activities are relatively better than elsewhere.

Irrigation is marginal in the zone, owing mainly to the favourable climate conditions for rainfed and market opportunities. Water resources are abundant in most places. Therefore, possibilities for water-based interventions are relatively marginal.

Highland temperate zone

This zones covers 440 000 km² (2 percent of the area of the region). Ten million ha of cultivated land (4 percent of the regional total) support a rural population of 30 million (7 percent of the regional total). This zone is located mainly in the Ethiopian and Eritrean highlands at an altitude of 1 800–3 000 m, and the climate is predominantly subhumid or humid. Given the high altitude, this zone is typically monomodal, and presents one single and long growing season. Temperate cereals, such as wheat, teff (in Ethiopia) and barley, are the most common sources of livelihood, complemented with pulses and potatoes. Livestock are relatively abundant and an important source of cash. Some households have access to soldiers' salaries (Ethiopia and Eritrea) or remittances (Lesotho), but these mountain areas offer few local opportunities for off-farm employment.

The particular agroclimatic conditions of the zone have a twofold effect on its rural livelihood conditions. On the one hand, the population is highly vulnerable owing to the early and late frosts at high altitudes that can severely reduce yields, and crop failures are not uncommon in cold and wet years. On the other hand, there is a considerable potential for diversification into higher-value temperate crops. The potential exists for substantial increases in agricultural productivity through a combination of water and soil-fertility-related interventions, in particular through better soil moisture management and small-scale irrigation.

Highland perennial zone

This relatively small livelihood zone is located mainly in the highlands of East African, covering an area of about 320 000 km² (1 percent of the regional total). The climate is mostly subhumid or humid, with an average growing period of more than 250 days. The rural population is 32 million (8 percent of the regional total). This zone has the highest population density in the region (more than 1 inhabitant/ ha). Therefore, the pressure on land is intense, and about 7 million ha of land are cultivated, mainly by smallholders. The average cultivated area per household is slightly less than 1 ha, but more than 50 percent of holdings are smaller than 0.5 ha. The livelihood base of this zone is characterized by perennial crops such as banana, plantain, enset, coffee and cassava, complemented by annual root crops, such as sweet potato and yam as well as pulses and cereals. Given the limited availability of pastures, livestock are a minor resource, amounting to about 6.2 million head. The main trends are diminishing farm size, declining soil fertility, and increasing poverty and hunger. People cope by working the land more intensively, but returns to labour are low.

Given the favourable conditions for rainfed agriculture, irrigation is a minor practice and accounts for only 52 000 ha (1 percent of the regional total). However, in conditions of heavy pressure on land resources, there is some scope for intensification through improved water control.

Tree crop zone

This zone is located in the Gulf of Guinea, with smaller pockets in the Democratic Republic of the Congo and Angola, largely in the humid zone. The zone occupies about 730 000 km² (3 percent of the regional total), accounts for 14 million ha of cultivated land (6 percent of the regional total), and is home to a rural population of almost 30 million (7 percent of the regional total). The production base of the zone is industrial tree crops, particularly cocoa, coffee, oil palm and rubber. Food crops are intercropped with tree crops and are grown mainly for self-consumption. Livestock are marginal (2 percent of the regional total). There are also commercial tree crop estates (particularly for oil palm and rubber), providing some employment opportunities for smallholder tree crop farmers through nucleus estate and outgrower schemes. As neither tree crop nor food crop failure is common, price fluctuations for industrial crops constitute the main source of vulnerability.

Given the favourable climate, irrigation is marginal in the region, and prospects for livelihood enhancement through water intervention are minor.

Forest-based zone

This zone occupies 2.6 million km² (11 percent of the total land in the region), accounts for 11 million ha of cultivated area (5 percent of the regional total), and is home to a rural population of 29 million (7 percent of the regional total). Most of the land lies in the humid forest zone of the Democratic Republic of the Congo. Farmers practise shifting cultivation, clearing new fields from the forest every year, cropping it for 2–5 years (cereals or groundnuts, followed by cassava) and then abandoning it to bush fallow for 7–20 years. Cassava is the main staple, complemented by maize, sorghum, beans and cocoyams. Sources of food and cash, in limited part, are also forest products and wild game. The livestock population is 3.2 million head (2 percent of the regional total), as pastoral land is limited, given the prevalence of forest vegetation. Rural infrastructures are poorly developed and access to markets is restricted. This implies agriculture of a largely subsistence nature.

While the irrigation potential (6.7 million ha) and the internal renewable water resources (1 460 km³/year) are the highest in the region, irrigation is marginal (87 000 ha) and represents 1 percent of the regional total. This zone offers little prospect for water-based interventions in support of poverty reduction in rural areas.

Large commercial and smallholder zone

This zone covers almost the whole of South Africa and the southern part of Namibia, Zambia and Zimbabwe. The climate is mostly semi-arid. The zone covers 1.23 million km² (5 percent of the regional total), with 15 million ha of cultivated land (7 percent of the regional total). It is home to 20 million rural people (5 percent of the regional total). It comprises two distinct types of farms: scattered smallholder farming in the homelands; and large-scale commercial farms. Both types are largely mixed cereal–livestock zones, with maize dominating in the north and east, and sorghum and millet in the west. Ruminants are abundant in this zone, but the level of crop–livestock integration is limited.

Irrigation is extensively used and has reached its full potential in many places, leading to competition for water between farmers and between sectors. Together with highly intense farming, irrigation is depriving soils, and the zone is becoming more drought-prone. In this zone, water-related interventions should concentrate on water productivity increases through improved management of agricultural water, and the development of water harvesting to support supplementary irrigation. Institutional issues, including issues of water rights, conflict resolution and river basin management, deserve particular attention.

Rice–tree crop zone

This zone is located exclusively in Madagascar – and benefits from a moist subhumid climate. It is the smallest zone of the region, accounting for less than 310 000 km² (1 percent of the regional total), of which 2.7 million ha are cultivated (1 percent of the regional total). The rural population is 8 million (2 percent of the regional total). Banana and coffee cultivation is complemented by rice, maize, cassava and legumes. Livestock are almost insignificant (about 1 million head).

Farms are small, and there is a significant amount of basin flood irrigation – equivalent to 10 percent of the total irrigated area of the region – used almost exclusively for paddy rice production, the main staple food in Madagascar. As irrigation is reaching its full potential in places, there is ample scope for increased productivity of irrigated agriculture through better water management.

Coastal artisanal fishing zone

This zone stretches all around the coastal areas of SSA. The zone covers 380 000 km^2 (2 percent of the regional total). It is home to accounts for 15.5 million rural people (4 percent of the regional total); most of the population of this zone live in urban areas (73 percent). People's livelihoods are based on artisanal fishing supplemented by crop production, sometimes in multistoried tree crop gardens with root crops under coconuts, fruit trees and cashews, plus some animal production. The cultivated land area of 3.6 million ha is only 2 percent of the regional total. Livestock numbers are small (fewer than 2 million head, or 1 percent of the regional total).

Irrigation is not very developed – 300 000 ha (4 percent of the regional total). However, as the coastal area has a high concentration of urban population, good prospects exist for the development of peri-urban agriculture, in which water control plays an important role. Therefore, in places, and according to market conditions, this zone offers prospects for further irrigation development.

Other relevant local zones
Peri-urban zone

Urban centres usually offer opportunities for rural people in terms of markets for farm products and labour. Agriculture areas around cities are characteristically focused on horticultural, livestock production, and off-farm work. Within the estimated total urban population of more than 200 million in the region, there is a significant number of farmers in cities and large towns. In some cities, it is estimated that 10 percent or more of the population are engaged in peri-urban agriculture. Overall, there are about 11 million agricultural producers in peri-urban areas. This livelihood zone is very heterogeneous, ranging from small-scale, capital-intensive, market-oriented vegetable-growing, dairy farming and livestock fattening, to part-time farming by the urban poor to cover part of their subsistence requirements. The level of crop–livestock integration is often low, and there are typically environmental and food quality concerns associated with peri-urban farming. The potential for poverty reduction is relatively low, mainly because the absolute number of poor is low. Agricultural growth is likely to take place spontaneously, in response to urban market demand for fresh produce, even in the absence of public-sector support. Unless curbed by concerns over negative environmental effects, rapid adoption of improved technologies can be expected. Overall, this is a dynamic livelihood zone with considerable growth potential.

Irrigated zone

Irrigated areas are scattered across the region, and they provide a broad range of food and cash crops, including rice, vegetables, cotton, and sugar cane. Irrigation constitutes a special case in relation to the heterogeneity of livelihood zones. Where irrigation-based production is the principal source of livelihood in an area, as in the case of large-scale irrigation schemes, the entire area can be considered an irrigation-based livelihood zone. Water control may be full or partial. Irrigated holdings vary considerably in size. Water shortages, deterioration of infrastructure, and reduced margins for main irrigated products are among the main problems facing farmers in irrigated areas. Many state-run schemes are currently in financial crisis, but if institutional and market problems can be solved, prospects for future agricultural growth are good. The incidence of poverty is lower than in other livelihood zones, and the absolute numbers of poor are small.

ANNEX 2
METHOD FOR ASSESSING INVESTMENT POTENTIAL

This annex describes the method used to assess the potential for investments in SSA. It also shows the potential outcomes, in table form, by livelihood zone and type of intervention. In order to determine priority for action in the different livelihoods zones, the method utilized the following three criteria:

o prevalence of poverty;

o water as a limiting factor for rural livelihoods;

o potential for water intervention.

The steps followed in order to generate the assessment are described below.

Step 1: Quantifying priorities according to the three criteria

This entailed a quantification of the three priority levels (low, moderate and high) for the criteria used in the analysis (above). Coefficients were applied to represent these three levels as a percentage of possible interventions for the criteria related to water as a limiting factor and poverty incidence: 100, 50 and 15 percent. The criterion relating to potential for intervention was based on population, land and water data (Table A2.1).

Step 2: Assessing unit costs by type of intervention

Costs have been assessed on the basis of data available at FAO from a large number of investment projects in the region. In view of the wide range of possible interventions and associated costs, such an assessment can only be viewed as a very rough estimate of such a potential for action and associated costs. Unit costs related to irrigation and land improvement are relatively well known. Costs of multiple-use systems have been assessed on the basis of a recent study (Renwick *et al.*, 2007), considering one system per household. The two types of interventions for which unit cost estimates are most difficult are those related to livestock watering and small-scale water harvesting infrastructures. For water harvesting, the costs associated with the range of possible technical options makes any assessment of an "average" cost very difficult. In order to be able to compare the different technologies, water harvesting interventions were expressed per unit of volume stored. A value of USD1/m³ was chosen. Table A2.2 shows the unit costs selected for this assessment. In view of the uncertainty associated with these costs, no attempt was made to differentiate between the livelihoods zones.

TABLE A2.1

Weighting factor for priority for action by livelihood zone

Livelihood zone	Poverty incidence	Water as limiting factor	Potential for water interventions
Arid	15	100	
Pastoral	100	100	
Agropastoral	100	100	
Cereal-based	100	100	
Cereal-root crop	100	100	
Root-crop-based	50	15	
Highland Temperate	100	75	
Highland Perennial	50	50	Based on population, land and water data
Tree crop	15	15	
Forest-based	50	15	
Large Commercial and Smallholder	15	100	
Rice-tree crop	50	15	
Coastal Artisanal Fishing	15	15	

TABLE A2.2

Unit costs, USD/unit

Manage soil moisture in rainfed areas	Invest in small-scale water harvesting infrastructure	Promote small-scale community-based irrigation	Improve existing irrigation systems	Improve water control for peri-urban producers	Invest in water for livestock production	Facilitate multiple use of water
ha	mm³	ha	ha	ha	head	household
75	1 000 000	4 250	2 000	3 000	30	75

Step 3: Assessment of the "absolute" potential for interventions by livelihood zone

The absolute potential for each intervention by livelihood zone represents the maximum possible extent of each type of intervention in each zone, irrespective of the role of water as a limiting factor and of the incidence of poverty in the area. The results are presented in Table A2.3. The potential was assessed on the basis of demographic and natural resources as follows:

- Manage soil moisture in rainfed areas: Extent of rainfed cultivated land in the zone (unit: ha).
- Small-scale water harvesting: the lower of the following two: (i) 80 percent of local runoff (considering a 20-percent "environmental" flow); or (ii) 30 percent of the rainfed cultivated land multiplied by 1 000 m^3/ha (unit: million m^3).
- Small-scale community-based irrigation: the lower of the following two: (i) current extent of small-scale irrigation (i.e. this would correspond to a doubling of existing small-scale irrigation infrastructure); or (ii) the difference between potential irrigation and actual irrigation (unit: ha).
- Improve existing irrigation systems: 50 percent of existing irrigation.
- Water control for peri-urban producers: 0.008 ha per inhabitant in urban areas, based on assessment made in Ghana (unit: ha).
- Water for livestock production: number of livestock (cattle) in the livelihood zone (unit: head).
- Multiple use of water: number of rural households in the zone, with an estimated 5 persons per household (unit: household).

TABLE A2.3

Absolute potential

Livelihood zone	Manage soil moisture in rainfed areas	Invest in small-scale water harvesting infrastructure	Promote small-scale community-based irrigation	Improve existing irrigation systems	Improve water control for peri-urban producers	Invest in water for livestock production	Facilitate multiple use of water
	ha	Mm³	ha	ha	ha	head	household
Arid	765 135	230	200 000	389 793	62 606	8 368 400	1 668 478
Pastoral	8 948 023	2 684	500 000	601 019	113 497	24 223 700	5 448 920
Agropastoral	41 547 366	12 464	600 000	458 437	234 625	35 174 400	7 686 340
Cereal-based	35 413 458	10 624	499 407	312 130	322 533	24 497 200	13 180 280
Cereal-root crop	51 176 547	15 353	358 122	223 826	249 844	38 576 100	13 588 440
Root-crop-based	28 619 812	8 586	149 226	93 267	222 446	16 240 100	9 742 340
Highland Temperate	10 101 891	3 031	138 838	86 774	123 970	12 377 500	6 006 700
Highland Perennial	7 026 607	2 108	43 088	26 930	107 556	6 254 820	6 551 020
Tree crop	13 567 324	4 070	92 743	57 965	189 631	4 186 170	5 924 960
Forest-based	10 915 013	3 275	73 212	45 758	147 982	3 327 710	5 834 060
Large Commercial and Smallholder	13 849 601	4 155	0	709 010	118 778	12 833 100	4 087 820
Rice-tree crop	2 007 666	602	86 686	346 763	30 521	1 153 460	1 610 470
Coastal Artisanal Fishing	3 257 752	977	298 859	186 787	206 410	1 967 010	3 111 620
Total	227 196 195	68 159	3 040 181	3 538 456	2 130 401	189 179 670	84 441 448

Step 4: Assessment of the intervention potential

The intervention potential was calculated by applying the coefficients of Table A2.1 to each combination of intervention and livelihood zone. The coefficients were modified for poverty incidence in three cases. In the cases of irrigation improvement and peri-urban producers, no reduction coefficient was applied. In the case of multiple-use systems, it was estimated that the need for multiple-use systems could never be more than 90 percent of the households.

Step 5: Assessing the number of people reached for each intervention

For soil moisture management and small-scale water harvesting, the number of persons per hectare and per 1 000 m³ of water respectively was estimated by multiplying the number of rural people in the zone by a coefficient representing the number of crop farmers, and dividing by the rainfed cultivated area in the zone. For small-scale irrigation, improvement in irrigated systems and peri-urban producers, the area was multiplied by the average number of farmers per hectare (estimated at 10 farmers per hectare). Livestock was calculated by dividing the number of head by the rural population, and multiplying by a coefficient representing the percentage of households having animals. Multiple-use systems were calculated considering 5 persons per household. These figures are summarized in Table A2.4.

Step 6: Calculating investment costs

The investment costs were calculated by multiplying the relevant intervention figures of the livelihood zones by the unit costs of Table A2.2.

TABLE A2.4

Number of people reached per unit

Livelihood zone	Manage soil moisture in rainfed areas	Invest in small-scale water harvesting infrastructure	Promote small-scale community-based irrigation	Improve existing irrigation systems	Improve water control for peri-urban producers	Invest in water for livestock production	Facilitate multiple use of water
	pers./ha	Mm3	ha	ha	ha	head	household
Arid	0.54	540	10	10	10	0.90	5
Pastoral	0.27	268	10	10	10	1.01	5
Agropastoral	0.45	452	10	10	10	0.87	5
Cereal-based	1.46	1462	10	10	10	1.35	5
Cereal-root crop	1.05	1052	10	10	10	0.88	5
Root-crop-based	1.35	1352	10	10	10	1.20	5
Highland Temperate	2.34	2338	10	10	10	0.97	5
Highland Perennial	3.70	3700	10	10	10	2.09	5
Tree crop	1.73	1732	10	10	10	2.83	5
Forest-based	1.86	1855	10	10	10	3.51	5
Large Commercial and Smallholder	0.94	937	10	10	10	0.80	5
Rice-tree crop	2.38	2384	10	10	10	2.79	5
Coastal Artisanal Fishing	2.14	2142	10	10	10	3.16	5

CLIMATE CHANGE IMPLICATIONS
FOR AGRICULTURE
IN SUB-SAHARAN AFRICA

Lim Li Ching

CONTENTS

BOXES

ABOUT THE AUTHOR

Lim Li Ching is a researcher with the Third World Network. She was a lead author contributing to the East and South Asia and Pacific regional report of the International Assessment of Agricultural Knowledge, Science and Technology for Development (IAASTD).

INTRODUCTION

One sixth of humanity is suffering from hunger – the Food and Agriculture Organization of the United Nations (FAO) estimates that 105 million more people were pushed into hunger in 2009, bringing the total number of hungry to a shameful 1.02 billion.

Add climate change to these statistics, and the situation becomes even more urgent. A recent report by the International Food Policy Research Institute (IFPRI) warns that unchecked climate change will have major negative effects on agricultural productivity, with yield declines for the most important crops and additional price increases for the world's staples – rice, wheat, maize and soybeans (Nelson *et al.*, 2009). It also suggests that there will be 20 percent more malnourished children in 2050 due to climate change.

According to the International Assessment of Agricultural Knowledge, Science and Technology for Development (IAASTD, 2009)[1], climate change, coincident with increasing demand for food, feed, fibre and fuel, has the potential to irreversibly damage the natural resource base on which agriculture depends, with significant consequences for food insecurity. Climate change could also significantly constrain economic development in developing countries that rely largely on agriculture (Rosegrant *et al.*, 2008).

---

1 The IAASTD is the most recent and comprehensive assessment of agriculture, co-sponsored by the World Bank, FAO, United Nations Environment Programme (UNEP), United Nations Development Prgoramme (UNDP), World Health Organization (WHO), United Nations Educational, Scientific and Cultural Organization (UNESCO) and the Global Environment Facility (GEF).

CLIMATE CHANGE AND AGRICULTURE
Impacts on agricultural production

The Intergovernmental Panel on Climate Change (IPCC) has projected that crop productivity would increase slightly at mid- to high-latitudes for local mean temperature increases of up to 1–3°C (depending on the crop), but further warming would have increasingly negative impacts in all regions (Easterling *et al.*, 2007; IPCC, 2007a). More significantly, for many developing countries, at lower latitudes, especially in the seasonally dry and tropical regions, crop productivity is projected to decrease for even small local temperature increases (1–2°C).

Moreover, it is changes in the frequency and severity of extreme climate events, such as droughts and heavy precipitation, which will have more serious consequences. Modelling suggests that increasing frequency of crop loss due to such extreme climate events may overcome any positive effects of moderate temperature increase (Easterling *et al.*, 2007). Increases in the frequency of droughts and floods are projected to affect local crop production negatively, especially in subsistence sectors at low latitudes (IPCC, 2007a).

On a global scale, the potential for food production is projected to increase with increases in local average temperature over a range of 1–3°C, but above this it is projected to decrease (IPCC, 2007a). This would increase the number of people at risk of hunger. Climate change alone is estimated to increase the number of undernourished people to between 40 million and 170 million, although impacts may be mitigated by socio-economic development (Easterling *et al.*, 2007).

For sub-Saharan Africa, including where some of the poorest people live and farm, the projections of climate change's impacts on agriculture are dire (see Box 1). New studies confirm that Africa is one of the most vulnerable continents to climate variability and change because of multiple stresses and low adaptive capacity (IPCC, 2007a).

BOX 1
IMPACTS OF CLIMATE CHANGE ON AGRICULTURE IN SUB-SAHARAN AFRICA

Agricultural production – including access to food – in many African countries and regions will be severely affected. The area suitable for agriculture, the length of growing seasons and yield potential, particularly along the margins of semi-arid and arid areas, are expected to decrease. This would further adversely affect food security and exacerbate malnutrition. In some countries, yields from rain-fed agriculture, which is important for the poorest farmers, could be reduced by up to 50 percent by 2020.

Yields of grains and other crops could decrease substantially across the continent because of increased frequency of drought, even if there are potential production increases due to increases in carbon dioxide concentrations. Some crops (e.g. maize) could be discontinued in some areas. Livestock production would suffer due to deteriorated rangeland quality and degradation of rangeland areas.

There is evidence that freshwater resources, on which the viability of agriculture depends, are vulnerable and will be strongly impacted by climate change, and that current water management practices may not be sufficient to cope with these impacts. By 2020, between 75 and 250 million Africans could be exposed to increased water stress. Coupled with increased demand, this will adversely affect livelihoods and exacerbate water-related problems.

As a consequence of these impacts, climate change is likely to further entrench food insecurity in sub-Saharan Africa. By 2080, about 75 percent of all people at risk of hunger are estimated to live in this region.

Sources: Easterling *et al.*, 2007; IPCC, 2007a & 2007b

Who are the most vulnerable?

It is the majority of the world's rural poor who live in areas that are resource-poor, highly heterogeneous and risk-prone, who will be hardest hit by climate change. In particular, smallholder and subsistence farmers, pastoralists and artisanal fisherfolk will suffer complex, localised impacts of climate change and will be disproportionately affected by extreme climate events (Easterling *et al.*, 2007). For these vulnerable groups, even minor changes in climate can have disastrous impacts on their lives and livelihoods (Altieri and Koohafkan, 2008).

With a large number of smallholder and subsistence farming households in the dryland tropics, there is particular concern over temperature-induced declines in crop yields, and increasing frequency and severity of drought, which will lead to the following general impacts: increased likelihood of crop failure; increased diseases and mortality of livestock and/or forced sale of livestock at disadvantageous prices; increased livelihood insecurity, resulting in sale of other assets, indebtedness, out-migration and dependency on food aid; and a downward spiral on human development indicators such as health and education (Easterling *et al.*, 2007).

Other physical impacts of climate change important to smallholders are decreased water supply for major smallholder irrigation systems, effects of sea level rise on coastal areas, increased frequency of landfall tropical storms and other forms of environmental impact such as remobilisation of dunes in semi-arid southern Africa. There will also be impacts on human health, such as increased malaria risk affecting labour available for agriculture and other non-farm rural economic activities.

Agriculture's contribution to climate change

The relationship between climate change and agriculture is however a two-way one; climate change in general adversely affects agriculture and agriculture contributes to climate change in several major ways.

Agriculture directly releases into the atmosphere a significant amount of carbon dioxide (CO_2), methane (CH_4) and nitrous oxide (N_2O), amounting to around 10–12 percent of global anthropogenic greenhouse gas emissions annually (Smith *et al.*, 2007). Of global anthropogenic emissions in 2005, agriculture accounted for about 60 percent of nitrous oxide and about 50 percent of methane, both of which have far greater global warming impact than carbon dioxide. Nitrous oxide emissions are mainly associated with nitrogen fertilizers and manure applied to soils, as fertilizers are often applied in excess and not fully used by crops, so that some of the surplus is lost to the atmosphere. Fermentative digestion by ruminant livestock contributes to methane emissions, as does cultivation of rice in flooded conditions.

If indirect contributions (e.g. land conversion to agriculture, fertilizer production and distribution and farm operations) are factored in, it is estimated that the contribution of agriculture could be as high as 17–32 percent of global anthropogenic emissions (Bellarby *et al.*, 2008). In particular, land use change, driven by industrial production methods, accounts for more than half of total agricultural emissions.

Conventional industrial agriculture is also heavily reliant on fossil fuels. The manufacture and distribution of synthetic fertilizers contributes a significant amount of greenhouse gas emissions, between 0.6–1.2 percent of the world's total emissions (Bellarby *et al.*, 2008). This is because the production of fertilizers is energy intensive and emits carbon dioxide, while nitrate production also generates nitrous oxide.

While sub-Saharan Africa is a region where per capita food production is either in decline, or roughly constant at a level that is less than adequate, the rising wealth of urban populations is likely to increase demand, albeit slowly, for livestock products. This would result in intensification of agriculture and expansion to still largely unexploited areas, particularly in South and Central Africa (including Angola, Zambia, Democratic Republic of the Congo, Mozambique and Tanzania), with a consequent increase in greenhouse gas emissions. Hence, sub-Saharan Africa is, together with the Middle East and North Africa, projected to see the highest growth in emissions from agriculture, with a combined 95 percent increase in the period 1990 to 2020 (Smith *et al.*, 2007).

ECOLOGICAL AGRICULTURE HAS BOTH ADAPTATION AND MITIGATION POTENTIAL

The IAASTD clearly concluded that the "business as usual" scenario of industrial farming, input- and energy-intensiveness, along with damage to the environment is no longer tenable. In an era of climate change, the challenge is therefore to design an agriculture that adapts and responds to the changes in climate experienced, as well as reduces greenhouse gas emissions. This challenge could be met by ecological agriculture.

Sustainable and ecological agricultural approaches, including organic agriculture, can be in many forms, but generally integrate natural, regenerative processes; minimize non-renewable inputs (pesticides and fertilizers); rely on the knowledge and skills of farmers and depend on locally-adapted practices to innovate in the face of uncertainty (Pretty and Hine, 2001). Ecological agriculture is also biodiversity-based (Ensor, 2009), depending on and sustaining agricultural biodiversity.

Ecological agriculture is already practiced by many African smallholder farmers. Africa has approximately 33 million small farms, representing 80 percent of all farms in the region. Although Africa now imports large amounts of cereals, the majority of African farmers (many of them women) who are smallholders with farms below two hectares, produce a significant amount of basic food crops with virtually no or little use of fertilizers and improved seed (Altieri, 2008). They instead rely mainly on nature and natural processes, agricultural biodiversity, local resources and local knowledge to farm.

Any comprehensive strategy for addressing climate change must include both adaptation and mitigation (IFAD, 2008), and ecological agriculture has the potential to do both. For the most vulnerable people, whose livelihoods are being impacted now, adaptation is urgent. However, concerted and sustained mitigation efforts are also needed to prevent further deterioration in the medium term. Since adaptation becomes costlier and less effective as the magnitude of climate change increases, mitigation remains essential (Rosegrant *et al.*, 2008).

Ecological agriculture practices contribute to adaptation

Adaptation can be both autonomous and planned. Autonomous adaptation is the ongoing implementation of existing knowledge and technology in response to the changes in climate experienced; planned adaptation is the increase in adaptive capacity by mobilizing institutions and policies to establish or strengthen conditions that are favourable to effective adaptation, and investment in new technologies and infrastructure (Easterling *et al.*, 2007). Autonomous adaptation is highly relevant for smallholder farmers in developing countries (IFAD, 2008). Crucially, many of these autonomous adaptation options are met by ecological agriculture practices, as highlighted below. By increasing resilience within the agroecosystem, ecological agriculture increases its ability to continue functioning when faced with unexpected events such as climate change (Borron, 2006; Ensor, 2009).

Resiliency to climate disasters is closely linked to farm biodiversity; practices that enhance biodiversity allow farms to mimic natural ecological processes, enabling them to better respond to change and reduce risk. Thus, farmers who increase interspecific diversity suffer less damage compared to conventional farmers planting monocultures (Altieri and Koohafkan, 2008; Borron, 2006; Ensor, 2009; Niggli *et al.*, 2009). Moreover, the use of intraspecific diversity (different cultivars of the same crop) is insurance against future environmental change. Diverse agro-ecosystems can also adapt to new pests or increased pest numbers (Ensor, 2009).

Ecological agriculture practices that preserve soil fertility and maintain or increase organic matter – such as crop rotation, composting, green manures and cover crops – can reduce the negative effects of drought while increasing productivity (ITC and FiBL, 2007; Niggli *et al.*, 2009). In particular, the water holding capacity of soil is enhanced by practices that build organic matter, helping farmers withstand drought (Altieri and Koohafkan, 2008; Borron, 2006).

Conversely, organic matter also enhances water capture in soils, significantly reducing the risk of floods (ITC and FiBL, 2007; Niggli *et al.*, 2009). Practices such as crop residue retention, mulching, and agroforestry conserve soil moisture and protect crops against microclimate extremes.

In addition, water-harvesting practices allow farmers to rely on stored water during droughts, or to increase water availability. For example, in many parts of Burkina Faso and Mali there has been a revival of the old water harvesting system known as 'zai'. The zai are pits that farmers dig in rock-hard barren land, into which water would otherwise not penetrate. The pits are filled with organic matter and attract termites, which dig channels and thus improve soil structure so that more water can infiltrate and be held in the soil (Reij and Waters-Bayer, 2001, cited in Altieri and Koohafkan, 2008).

Adaptive capacity is an active process that involves the ability of individuals or communities to modify and transform practices in response to climate change. Indigenous and traditional knowledge are a key source of information on adaptive capacity, centred on the selective, experimental and resilient capabilities of farmers (Altieri and Koohafkan, 2008; Borron, 2006; IAASTD, 2009; ITC and FiBL; 2007, Niggli *et al.*, 2009). Many farmers cope with climate change, in various ways: minimising crop failure through increased use of drought-tolerant or disease- and pest-resistant local varieties, water-harvesting, extensive planting, mixed cropping, agroforesty, opportunistic weeding and wild plant gathering (see also Box 2).

BOX 2
EXAMPLES OF ADAPTATION STRATEGIES FROM SOUTHERN GHANA

Farmers in Southern Ghana are adapting and coping with climate variability in various ways. This is manifested by the diversity of resource management and cropping systems, which are based on indigenous knowledge of management of the fragile and variable environment, local genotypes of food crops, intercropping, and agroforestry systems. These coping mechanisms not only help meet farmers' subsistence needs, but also encourage biodiversity conservation. To offset crop failure arising from rainfall variability and unpredictability, farmers cultivate hardier (or drought-tolerant) types of the same crop species. The planting of vegetable crops that can serve as a hedge against risk associated with drought is also a common practice.

Source: Ofori Sarpong and Asante, 2004, cited in Altieri and Koofhafkan, 2008.

Ecological agriculture practices contribute to mitigation

On the other hand, agriculture has the potential to change from being one of the largest greenhouse gas emitters to a much smaller emitter and even a net carbon sink, while offering options for mitigation by reducing emissions and by sequestering carbon dioxide from the atmosphere into the soil. The solutions call for a shift to more sustainable farming practices that build up carbon in the soil and use less chemical fertilizers and pesticides (Bellarby *et al.* 2008; ITC and FiBL, 2007; Ziesemer, 2007).

There are a variety of practices that can reduce agriculture's contribution to climate change. These include crop rotations and improved farming system design, improved cropland management, improved nutrient and manure management, improved grazing-land and livestock management, maintaining fertile soils and restoration of degraded land, improved water and rice management, fertilizer management, land use change and agroforestry (Bellarby *et al.*, 2008; Niggli *et al.*, 2009; Smith *et al.*, 2007). Many of these practices are inherent in ecological agriculture and easily implemented.

In particular, it is estimated that a conversion to organic agriculture would considerably enhance the sequestration of carbon dioxide through the use of techniques that build up soil organic matter, as well as diminish nitrous oxide emissions due to no external mineral nitrogen input and more efficient nitrogen use (Niggli *et al.*, 2009). Organic systems have been found to sequester more carbon dioxide than conventional farms, while techniques that reduce soil erosion convert carbon losses into gains (Bellarby *et al.*, 2008; ITC and FiBL, 2007; Niggli *et al.*, 2009). Ecological agriculture is also self-sufficient in nitrogen due to recycling of manures from livestock and crop residues via composting, as well as planting of leguminous crops (Ensor, 2009; ITC and FiBL, 2007).

Moreover, practices rooted in ecological agriculture, such as introducing perennial crops to store carbon below ground and planting temporary vegetative cover between successive crops to reduce nitrous oxide emissions by extracting unused nitrogen, also mitigate climate change (Ensor, 2009).

CONCLUSION

Climate change will undoubtedly pose serious challenges for African agriculture. Nonetheless, African farmers are well poised to adopt ecological agriculture, given that many of these sustainable practices are already in place and contributing to food production. Ecological agriculture is also an option that is easily accessible for Africa, where many farmers cannot afford expensive chemical inputs.

Redesigning agriculture in an era of climate change would entail investing more resources, research and training into, providing appropriate policy support to, and implementing national, regional and international action plans on ecological agriculture. Doing so will not only be beneficial in terms of climate adaptation and mitigation, but will also be a paradigm shift towards increasing productivity while ensuring sustainability and meeting smallholder farmers' food security needs (IAASTD, 2009).

Ecological agriculture is indeed productive, as shown by a review of 286 projects in 57 countries, whereby farmers increased agricultural productivity by an average of 79 percent, after adopting 'resource-conserving' agriculture (Pretty *et al.*, 2006). A variety of sustainable technologies and practices were used, including integrated pest management, integrated nutrient management, conservation tillage, agroforestry, water harvesting in dryland areas, and livestock and aquaculture integration. The database was reanalysed to produce a summary of the impacts of organic and near-organic projects on agricultural productivity in Africa (Hine *et al.*, 2008). The average crop yield increase was even higher for these projects than the global average: 116 percent increase for all African projects and 128 percent increase for the projects in East Africa.

Maximising the synergies between adaptation and mitigation means that these strategies should be developed simultaneously. In particular, Khor (2009) suggests the following:

- There should be more research and action on adaptation measures in agriculture, especially in developing countries in order to assist farmers there to reduce the adverse impacts of climate change on agriculture.

- Action plans for mitigation measures for agriculture should be urgently researched and implemented.
- Financing assistance for adaptation and mitigation measures in the agriculture sector in developing countries should be prioritised.
- Arrangements should be made for the sharing of experiences and the transfer of good practices in agriculture that can constitute mitigation and adaptation.
- Given the many advantages of organic farming and ecological agriculture, in terms of climate change as well as social equity and farmers' livelihoods, there should be a much more significant share of research, personnel, investment, financing and overall support from governments and international agencies channelled towards ecological agriculture. Promotion of ecological agriculture can lead to a superior model of agriculture from the environmental and climate change perspective, as high-chemical and water-intensive agriculture is phased out, while more natural farming methods are phased in, with research and training programmes also promoting better production performances in ecological agriculture.

With appropriate focus on ecological agriculture as providing adaptation, mitigation and increased productivity options, a "win-win-win" scenario for agriculture is possible.

REFERENCES

Altieri, M.A. & P. Koohafkan. 2008. *Enduring farms: Climate change, smallholders and traditional farming communities.* Environment and Development Series No. 6. Third World Network, Penang.

Bellarby, J., B. Foereid, A. Hastings & P. Smith. 2008. *Cool farming: Climate impacts of agriculture and mitigation potential.* Greenpeace International, Amsterdam.

Borron, S. 2006. *Building resilience for an unpredictable future: How organic agriculture can help farmers adapt to climate change.* FAO, Rome.

Easterling, W.E., P.K. Aggarwal, P. Batima, K.M. Brander, L. Erda, S.M. Howden, A. Kirilenko, J. Morton, J.-F. Soussana, J. Schmidhuber & F.N. Tubiello. 2007. Food, fibre and forest products. p. 273-313. *In* M.L. Parry, O.F. Canziani, J.P. Palutikof, P.J. van der Linden and C.E. Hanson (eds.). *Climate Change 2007: Impacts, Adaptation and Vulnerability.* Contribution of Working Group II to the Fourth Assessment Report of the Intergovernmental Panel on Climate Change. Cambridge University Press, Cambridge, UK.

Ensor, J. 2009. Biodiverse agriculture for a changing climate. Practical Action, UK.

Hine, R., J. Pretty & S. Twarog. 2008. *Organic agriculture and food security in Africa.* UNCTAD and UNEP, Geneva and New York.

IAASTD. 2009. *Agriculture at a crossroads.* International Assessment of Agricultural Knowledge, Science and Technology for Development. Island Press, Washington D.C.

IFAD. 2008. Climate change and the future of smallholder agriculture. How can rural poor people be a part of the solution to climate change? Discussion paper prepared for the Round Table on Climate Change at the Thirty-first session of IFAD's Governing Council, 14 February 2008. International Fund for Agricultural Development, Rome.

IPCC. 2007a. Summary for Policymakers. p. 7-22. *In* M.L. Parry, O.F. Canziani, J.P. Palutikof, P.J. van der Linden and C.E. Hanson (eds.). *Climate Change 2007: Impacts, Adaptation and Vulnerability.* Contribution of Working Group II to the Fourth Assessment Report of the Intergovernmental Panel on Climate Change. Cambridge University Press, Cambridge, UK.

IPCC. 2007b. *Climate Change 2007: Synthesis Report.* Contribution of Working Groups I, II and III to the Fourth Assessment Report of the Intergovernmental Panel on Climate Change [Core Writing Team, Pachauri, R.K and Reisinger, A. (eds.)]. IPCC, Geneva, Switzerland.

ITC (International Trade Centre UNCTAD/WTO) and FiBL (Research Institute of Organic Agriculture). 2007. *Organic farming and climate change.* ITC, Geneva.

Khor, M. 2009. *Food crisis, climate change and the importance of sustainable agriculture.* Environment and Development Series No. 8. Third World Network, Penang.

Nelson, G.C., M.W. Rosegrant, J. Koo, R. Robertson, T. Sulser, T. Zhu, C. Ringler, S. Msangi, A. Palazzo, M. Batka, M. Magalhaes, R. Valmonte-Santos, M. Ewing & D. Lee. 2009. *Climate change. Impact on agriculture and costs of adaptation.* IFPRI, Washington D.C.

Niggli, U., A. Fließbach, P. Hepperly & N. Scialabba. 2009. *Low greenhouse gas agriculture: Mitigation and adaptation potential of sustainable farming systems.* April 2009, Rev. 2 – 2009. FAO, Rome.

Pretty, J. & R. Hine. 2001. *Reducing food poverty with sustainable agriculture: A summary of new evidence.* University of Essex Centre for Environment and Society, UK.

Pretty, J.N., A.D. Noble, D. Bossio, J. Dixon, R.E. Hine, F.W.T. Penning de Vries & J.I.L. Morison. 2006. Resource-conserving agriculture increases yields in developing countries. *Environmental Science and Technology (Policy Analysis)* 40(4): 1114-1119.

Rosegrant, M.W., M. Ewing, G. Yohe, I. Burton, S. Huq & R. Valmonte-Santos. 2008. *Climate change and agriculture: Threats and opportunities.* GTZ on behalf of Federal Ministry for Economic Cooperation and Development, Eschborn, Germany.

Smith, P., D. Martino, Z. Cai, D. Gwary, H. Janzen, P. Kumar, B. McCarl, S. Ogle, F. O'Mara, C. Rice, B. Scholes & O. Sirotenko. 2007. Agriculture. p. 497-540. *In* B. Metz, O.R. Davidson, P.R. Bosch, R. Dave and L.A. Meyer (eds.). *Climate Change 2007: Mitigation.* Contribution of Working Group III to the Fourth Assessment Report of the Intergovernmental Panel on Climate Change. Cambridge University Press, Cambridge, UK and New York, NY, USA.

Ziesemer, J. 2007. *Energy use in organic food systems.* FAO, Rome.

A MANUAL FOR THE PREPARATION OF WOREDA AND LOCAL COMMUNITY PLANS FOR ENVIRONMENTAL MANAGEMENT FOR SUSTAINABLE DEVELOPMENT

Developed by experts in the
Environmental Protection Authority of Ethiopia,
and translated from Amharic
with an introduction by
Tewolde Berhan Gebre Egziabher

CONTENTS

ABOUT THE AUTHOR

Tewolde Berhan Gebre Egziabher is the Director General of the Environmental Protection Authority of the Federal Democratic Republic of Ethiopia. He is a plant ecologist who has been a university academic staff member and an academic administrator as well as a negotiator on biodiversity issues.

EDITORIAL NOTES

What is written in this chapter about Ethiopia also applies to nearly all other countries, particularly those in sub-Saharan Africa.

'Woreda' is the lowest level of official administration in Ethiopia. It is approximately equivalent to a district in other countries. Below the woreda are 'kebeles', approximately equivalent to parishes, with between 150 and 300 households. Local communities are neighbourhood groups of usually around 40 households occupying one integrated area where their homesteads, cultivated fields, water resources and other natural resources are found.

INTRODUCTION

Environmental resources provide the goods and services needed for combating poverty and for economic growth. Therefore, they are the basis for social and economic development. However, owing to mismanagement and inappropriate utilization, their current contribution to Ethiopia's overall development is not as large as it could have been.

In addition to hindering development projects by reducing raw material availability, the problems with the management and use of environmental resources increase risks to diseases, floods, drought and desertification. When environmental resources are not properly managed, they deteriorate; and when access to their utilization is unjust, at least some of the members of the rural communities that depend on them become impoverished. When both equity and appropriate management are lacking, poverty becomes widespread and the rural people turn into environmentally unsustainable exploiters and refugees. If the process is not reversed through the speedy institution of equity and appropriate management, the problem can spiral out of control and engulf the whole country in crisis.

Environmental degradation thus reduces both personal and societal economic well-being. The environmental resources then get mined out. The land's capacity to provide biomass for feed and fuel drops. The hydrological cycle gets disrupted. Soil fertility reduces. Climate changes. Both the local ecosystem and the local community become unstable and attempts to reduce poverty and bring about development become ineffective. For these reasons, reversing environmental degradation and eradicating poverty are mutually supportive. Therefore, they both need to be jointly undertaken in Ethiopia's attempts to develop. Ethiopia's national economic development programmes, national environmental management systems and international environmental obligations must thus be harmonized in implementation for a self-magnifying synergistic impact.

This means that Ethiopia has to do all that it can to develop socially and economically by sustainably using its existing natural and man-made as well as cultural resources so as to improve the well-being of its citizens without reducing the opportunities of its future generations to maintain their well-being.

That is why Ethiopia's vision states: "To see Ethiopia become a country where democratic rule, good-governance and social justice reigns, upon the involvement and free will of its peoples; and once extricating itself from poverty becomes a middle-income economy".

Strategic objectives

The strategic objectives of environmental management for sustainable development are thus to:

- Ensure gender equality and motivate local community initiatives to strengthen environmental protection and the sustainable use of natural resources so as to improve life for each and all;
- Rehabilitate degraded ecosystems to their original productive capacity especially in food, feed and biomass for fuel and for maintaining biodiversity;
- Increase the ability of the environment for the sustainable availability of raw materials and services for development;
- Eliminate the harm to health that comes from urban waste;
- Protect the environment from chemical pollution; and
- Ensure that the dictates of environmental well-being and the observance of human values including equity, especially gender equality, are incorporated in all the processes of economic development.

Implementation strategies

In order to use environmental resources while ensuring equity, gender equality and people's participation, the implementation strategies are to:

- Foster the participation of all stakeholders in the management and use of environmental resources;

○ Apply techniques that improve the management of natural resources and the efficiency of their sustainable use;

○ Establish, to this effect, systems for dissemination of environmental information;

○ Generate support and carry out activities needed to remove impediments to the preparation and implementation of plans for environmental management for sustainable development by the rural local communities themselves;

○ Improve capacity on a continuing basis starting with what is already existing in the organized local communities and incrementally adding what becomes needed as time goes on; and

○ Develop and implement to this effect awareness raising and information dissemination programmes for improving capacity for the management and sustainable use of environmental resources by all sectors of society.

WHY PLANS DRAWN UP BY WOREDAS AND LOCAL COMMUNITIES FOR ENVIRONMENTAL MANAGEMENT FOR SUSTAINABLE DEVELOPMENT ARE NEEDED

The full participation in, and the ownership of, the processes for the preparation and implementation plans for environmental resources management for sustainable development by all the members of local communities ensure the inclusion of everybody concerned at all levels of decision taking.

The lowest unit of government administration is the woreda (district). Therefore, the harmonization of economic and social development initiatives with the management and maximization of the sustainable use of environmental resources can best be effected at this level. This is because the complete participation of the population in the process is easiest to achieve at the local community level. Therefore, it makes sense to synthesize even the woreda level plans for environmental management for sustainable development out of its constituent local community level plans.

The respective woreda and local community plans for environmental management for sustainable development should internalize the objective conditions of the local environmental resources and undertake to improve them through the use of the knowledge, skills and labour of the local communities, realistically augmented by additional scientific knowledge and skills new to them. To this effect, all stakeholders should participate in the preparation and implementation of the respective plans for environmental management for sustainable development.

The plans for environmental management for sustainable development can thus effectively implement the land and biodiversity provisions of the Ethiopian Environmental Policy and the environmental conventions to which Ethiopia is a party (the Convention to Combat Desertification, the Convention on Biological Diversity, the Framework Convention on Climate Change etc.)

Effective environmental management for sustainable development plans can then combat Ethiopia's problem of environmental resources degradation and contribute by 2015 to the reduction by half of Ethiopians that suffer from extreme poverty, as required by the first of the United Nations Millennium Development Goals. It is towards this end that one of the aims of the Environmental Protection Authority of Ethiopia is the development and implementation of plans for environmental management for sustainable development by 125 woredas of the country over the next two years, i.e. from 2009 to 2010.

The local communities in these woredas will be helped to strengthen their respective environmental management institutions and thus also their respective capacities to generate wealth and save money, to identify and solve their environmental problems and to use their environmental resources as new sources of income while at the same time reducing their vulnerability to human impacts.

Definitions

In this document, the following terms will have the meanings respectively specified:

Local community means the farming families who live in a village or in neighbouring villages in a specified kebele and have agreed to jointly manage, protect and use their natural resources so as to jointly combat land degradation and droughts as well as floods through coordinating their respective activities and thus care for their whole ecosystem.

Sustainable development means achieving through mutual cooperation and gender equity and equality economic improvement for equity and self-sufficiency of each local community member without endangering the environmental well-being for, and economic opportunities of, future generations.

Local community environmental management for sustainable development means the participatory discussion by the members of the local community to clearly define and carry out through jointly providing the required financial resources, implements and labour for the activities aimed at achieving a development that will continue uninterrupted into the future.

THE PREPARATION OF A LOCAL COMMUNITY PLAN FOR ENVIRONMENTAL MANAGEMENT FOR SUSTAINABLE DEVELOPMENT

The plan for environmental management for sustainable development of a woreda is derived from the compilation into a single plan of the constituent local community plans. The woreda level compilation is then implemented by the woreda administration and the respective local communities with the support of other stakeholders as and when available and required.

Planning stages

o Step 1: Choosing woredas

Regional (federal administrative units) environmental protection offices use their respective criteria to identify the woredas that should be given the highest priority for support to develop and implement their plans for environmental management for sustainable development. The Environmental Protection Authority has preliminarily identified criteria (see below) to help the regional environmental protection offices in defining their respective criteria.

o Step 2: Determination of the vision for the environmental management for sustainable development of a selected woreda

Based on Ethiopia's development vision and on an evaluation of the objective realities of the environment and development situation of the constituent local communities, a woreda administration with the participation of its local communities and available stakeholders puts into words the woreda's vision of environmental management for sustainable development.

o Step 3: Determination of the woreda objectives for environmental management for sustainable development

The woreda objectives for environmental management for sustainable development are also articulated through the participation of the constituent local communities as well as available stakeholders taking into account the condition of the existing natural resources and the social, economic and ecological dimensions of the constituent local communities.

o Step 4: The preparation and implementation of local community plans for environmental management for sustainable development

Because it has to be the local community's instrument for enabling the rural people themselves to bring about their own development, the plan for environmental management for sustainable development of each local community has not only to be implemented but also to be developed by the respective local community through the full participation and ownership of all its members.

The local community through the full participation of its membership thus enunciates the targets it wants to reach. It then specifies the sources and quantities of the financial and other resources, implements and labour required to reach each of the targets.

Finally, it determines what is to be done when and by whom, and this produces the local community's plan for environmental management for sustainable development, which the local community implements.

Preparation of the local community plan
Selecting local communities

Through the participation of all the kebele farmers' associations, the woreda administration works to mobilize all the local communities (villages) in each kebele together, in groups or individually, as they see fit, to prepare their respective plans for environmental management for sustainable development. To this effect, the woreda administration assigns the work of mobilization to agricultural extension agents or to other experts, as it sees appropriate.

The initial local community plans for environmental management for sustainable development prepared in a woreda will obviously be used by the woreda for learning how the process of planning and implementation is best promoted. The first local communities that will prepare their plans should thus be selected with care for their appropriateness in providing experience that is likely to be valuable throughout, or at least through large parts of, the woreda in both the planning and implementation phases.

Awareness building

Both the members of a local community and all stakeholders need to know about the need for and the specifics of the planning process. They must thus also know the nature of the current environmental resources management within the area occupied by the local community and the objective of the planning process that is to be initiated. It is only then that they can all come to an agreement on what is to be done and how best to do it.

Information gathering

Gathering information on the local environment and discussing it to reach a common appreciation of its condition is a necessary first step. The local community then discusses the causes of the problems that it identifies, and evaluates the strengths and weaknesses both in itself and in the environment so as to be able to decide how to solve those problems and to rehabilitate the environment effectively.

In this way, the local community evaluates the environmental resources that are available, their adequacy to meet the future needs for its development, and the handicaps that it has in obtaining the needed resources. This information can lead it to the action required for effective environmental management and the improvement of human well-being.

Putting the local community vision into words

The local community thus becomes able to put into words its vision for environmental management for sustainable development based on its agreed analysis of the present condition of its environmental resources and an appreciation of what it realistically wants them to become.

Determining the objectives of the local community

The objectives determined by the local community are meant to show the work that must be done to bring about the content expressed in the vision for environmental management for sustainable development that has been put into words. These objectives will help generate a consensus on the strategic action that must be taken jointly by the local community and individually by its members to solve the identified problems. It will then become possible for the local community to agree on the time lines when the specific activities have to be carried out.

Setting the local community targets

After the objectives have been determined, the targets that have to be aimed at so as to achieve each objective are set. These targets must be specific enough to clearly show the action that must be taken to achieve them and the time, labour and resources that will be required. The outcomes must be measurable or at least capable of being described. The targets set will help assign responsibilities to specified local community members, and the measurement and description of the outcomes will enable the evaluation of achievements. Because their implementation requires the commitment of each individual that is to be involved, their specification has also to be made through a clear and exhaustive discussion that involves all of those who will implement them.

Specifying activities and apportioning responsibilities

Each of the activities to be carried out must be specified to show the problem that is to be solved, the time that it requires to carry out, the resources that must be made available to carry it out, the sources of each of the needed resource, and the practicality of its implementation.

Comparing and prioritizing activities

The local community develops agreed criteria and, using the criteria, prioritizes the identified activities. The criteria will specify the urgency of achieving the outcomes expected from each activity. Finally, the local community evaluates and accepts the prioritization of the activities.

Determining success indicators

The achieving of targets and thus the goals that gave rise to them must be evaluated by the local community itself so as to enable it to realize if the improvement in the environmental resources supply and thus also in human life that is aimed for is actually happening, and if that supply looks like it will continue into the future. Therefore, the local community needs to agree a set of criteria for making the evaluation. The criteria must be as concrete and as clear as possible and they must measure the level of maximization of benefits compared to expenditures in both resources and labour.

The participation of all members of the local community and stakeholders is essential to verify that the activities being planned do actually emanate from the objectives determined and that the targets set maximize the possibility of success as well as the accuracy of its evaluation through the use of the indicators that have been developed to measure it.

Finalizing the local community plan for environmental management for sustainable development

The plan for environmental management for sustainable development is formulated by the local community based on the consensus it reaches after considering all the ideas that its members and stakeholders that are directly or even indirectly involved in it express.

The substantive issues that should be considered during the planning process include the following:

- The implementability of each of the proposed activities by using the knowledge, skills, labour and resources found within the local community;
- Their potential in strengthening the community's knowledge, skills and natural resources;
- Their potential for continuing to help solve the local community's keenly felt problems;
- The extent to which the activities create productive employment;
- The extent to which they involve and benefit all the members of the local community;
- The clarity with which they specify who in the local community is to do what;
- The extent to which they have internalized the environmental and developmental conditions of the neighbouring local communities; and
- The extent to which they specify the problems and help to point to the solutions that will be encountered during implementation.

Implementation of the local community plan

If the local community agrees a set of bylaws that governs the implementation of its plan for environmental management for sustainable development, the coordination of action to be taken by each community member becomes easy.[1]

The preparation of the local community bylaws for environmental management for sustainable development

The bylaws for the implementation of the local community plan for environmental management for sustainable development are prepared by the local community

---

1   In Ethiopia, the role of local community bylaws in the environmental management for sustainable development is supported at all levels of government from the Prime Minister's Office through all administrative levels.

— 213 —

itself. The following draft merely aims at showing what it could look like. What is given here can thus act as a draft for a local community in starting to formulate its bylaws. Since nobody knows the internal conditions, collective wishes and existing capacity of any local community as well as the members of the local community itself, that local community can, and is encouraged, as it sees fit, to add to, subtract from or even altogether discard this draft to formulate its own bylaws afresh.

Recognition of the local community bylaws for environmental management for sustainable development

The plan for environmental management for sustainable development, especially the bylaws, thus formulated with the full participation of the members of the local community is first reviewed by the local community itself ensuring that the full participation of its members has been assured. Following the local community's approval, it is then submitted to the woreda administration. The woreda administration checks it, especially to ensure that the provisions of the bylaws formulated for implementing it do not violate the Constitution or any of the relevant national laws. The woreda administration then registers the bylaws, which become binding on the local community and all its members as well as any outsider who goes into the area under the management of the local community as of that date of registration.

DRAFT BYLAW FOR IMPLEMENTATION OF A LOCAL COMMUNITY PLAN FOR ENVIRONMENTAL MANAGEMENT FOR SUSTAINABLE DEVELOPMENT
Introduction

- Having decided to make our development sustainable by ensuring the compatibility of our social and economic development activities with the maximization of the well-being of our environment;

- Realizing that, by appropriately caring for the agricultural and natural biodiversity in our farms and on our uncultivated land and also by maximizing their production of biomass, we can sustainably increase our crop and animal production as well as satisfy our needs for wood for fuel and other purposes, while at the same time increasing water in the ground, in springs, and in streams;

- Convinced that environmental protection in both our private and communal land holdings will maximize success in the planning and implementation of our own development;

- We, the local community members of .... Village (Villages), ... Woreda, ...Region, have agreed, out of our own conviction and motivation, to care for and develop our environmental resources which are privately under the control of our respective families and communally under the control of our respective villages so that they satisfy our economic needs and improve the opportunities for our children to continue improving their well-being, and we have thus formulated these bylaws.

Name of our Association

The Association that we have created through these bylaws shall be called "... Local Community Environmental Management for Sustainable Development Association."

The geographical limits of the Association

Our Association adjoins .... on the east, ...on the west, ... on the north and .... on the south.

The objectives of the Association

- Every member of our Association who cultivates a piece of land shall prepare and apply compost to maintain the fertility of that piece of land at as high a level as possible.
- By planting trees on .... hectares of land each of us individually in and around his/her land holding and all of us communally in uncultivated community land, we shall produce enough wood for everyone of us to use as firewood or for other purposes in 2010 and beyond.
- Each of our members shall have grass and other forage plants on .... hectares, and shall thus have sufficient feed for his/her animals in 2014 and beyond.
- All the members of our Association shall join hands to construct terraces and bunds on .... hectares of sloping land and shall plant trees, shrubs and grass, as appropriate, to both protect the physical structures that we will have built and to make them more effective in preventing soil erosion.
- We shall close off .... hectares of land to all domestic animals and to people and we shall guard the closed off area to enable plants and wild animals to re-establish themselves, and shall develop a plan for the management of the closed off area.
- We shall create employment opportunities for .... people by the end of the year 2010 and continue creating more thereafter so as to increase alternative income opportunities and reduce pressure from the land.

Raising soil fertility

Each of our members shall:

- Undertake to raise soil fertility in, and to prevent soil loss from, his/her cultivated and uncultivated land holding, as appropriate, by digging and/or building soil conserving structures and by planting appropriate soil protecting and enriching species;
- Collect weeds, leaves, animal dung, chicken droppings, household wastes etc. and mix them to make compost that is sufficient in quality and quantity to fertilize his/her cultivated parcels of land;
- Participate or enable, as the case may be, his/her family members to participate, in training programmes organized for the purpose; and
- Plant fodder seedlings, especially shrubs and trees, around his/her homestead and parcels of cultivated land as well as on land that is not cultivated so as to create coordinated soil conservation and soil fertility increases according to the plan agreed by the Association.

Restricting domestic animal movement

- Every member shall participate in all meetings on the planning and implementation of the management of grazing land and animal production as well as in their evaluation organized by the Association.
- Each member shall participate in meetings and specified activities determined by the Association to stop free range grazing by domestic animals in a phased manner, to protect the area in which free range domestic animal grazing has been phased out, and to evaluate the impacts of the phasing out of, and protection from, free range domestic animal grazing on forage and other biomass production.
- Each member shall tether his/her animals or keep them in an enclosed space and cut and carry fodder to feed them.

Fodder production

Every member of the Association shall:

- Plant fodder plants around his/her parcel of cultivated land and his/her homestead, care for the fodder thus planted, and use the fodder appropriately;
- Harvest and store hay at the end of each rainy season, grow fodder during the dry season using rain water, irrigation or harvested water as the case may be, and collect and store chaff and other crop by-products so that his/her animals that have been prevented from free range grazing, especially milk cows and their calves, shall not suffer from feed shortage;
- Contribute his/her dues, as specified by the Executive Committee of the Association, in developing fodder on the communal land that has been closed off from free range animal grazing;
- Participate in the planning and implementation of strategies to counter falls in the selling price of domestic animals in times of drought and fodder shortage;
- Carry out the work assigned to him/her by the Association's Executive Committee to protect the area closed off from free range grazing; and
- Cut, carry and store at the time specified by the Association his/her share of fodder from the communal land that has been protected from free range animal grazing.

Meeting the needs for wood for fuel, construction and other purposes

- Each member shall grow and care for plants around his/her homestead and parcels of land for firewood, construction, shade, herbal medication, farming implements, handicrafts etc. as well as for fruits and fodder.
- Our Association shall allow trees, shrubs and tall grasses to grow as wind breaks and for maintaining soil fertility in areas that will not be farmed, including on

hilly terrain, and shall carry out enrichment planting as needed, and each of our members shall do the work assigned to him/her for the purpose.

o Each member shall contribute what is assigned to him/her to produce wood for fuel, construction, etc. on communal land and shall take his/her share from what has thus been communally produced.

The management and use of water

o Each member shall participate when appropriate techniques of harvesting and storing water as well as retaining soil moisture by the Association are discussed and determined, and shall individually or in a group, as determined by the Executive Committee of the Association, undertake the work assigned to him/her.

o When it is decided by the Association that specified areas around springs and along the edges of streams shall be free from human activity and from domestic animals, each member shall restrict the movements of his/her family members and domestic animals accordingly; the restricted area shall become accessible only for activities determined by the Association as appropriate.

Rehabilitating gullies and preventing gully formation

o Every member of our Association who has any gully formed or forming on his/her parcel of cultivable land has the obligation to prevent its expansion and to rehabilitate it.

o Gullies that are on uncultivated communal land and those that are on personal holdings but are seen by the Association as not capable of being rehabilitated by one farmer/household shall be rehabilitated by our Association. Every member of the Association shall participate in the gully treatment as assigned to him/her by the Executive Committee.

- Farmers with adjoining parcels of land that have been affected by a gully shall mutually agree on their respective contributions towards rehabilitating the gully and shall accordingly notify the Executive Committee, which shall then monitor the implementation of the agreed rehabilitation process.

The care and development of slopes

- We, the members of this Association, have agreed to rehabilitate the degraded hillsides in our village, which are .... hectares in extent and which are shown on the map that we have attached to this copy of our bylaws. To this effect, we have agreed to close off the indicated hillsides from access to domestic animals. Each of our members shall prevent his/her animals from going into this closed off area.
- The grass and the wood that grow in the closed off area shall be divided among us in a General Assembly of the Association for us to cut the grass and lop some branches of the trees and shrubs and carry away.

Qualification requirements for membership of our Association

- Any resident of our kebele (village, villages or sub-district) shall be a member of our Association. Any resident of our kebele (village, villages or sub-district) who does not want to be a member of the Association shall forfeit the right to graze domestic animals, to cut grass or to lop any branches of any shrub or tree from anywhere except his/her parcel of cultivable land.
- Each member of the Association shall contribute the labour, knowledge, skill or money that has been determined by the Executive Committee as his/her due to contribute.

Causes for expulsion from the Association

- Any member who refuses to obey these bylaws, who fails to be bound by the decisions of the General Assembly or the Executive Committee of the Association, or who fails to contribute with due diligence what has been specified to come from him/her for collective action, may be expelled from the Association by a decision of the Executive Committee.
- Any member who feels that any decision of the Executive Committee has been unfair can present his/her case to the General Assembly of the Association. The decision of the General Assembly shall be final.

The organizational structure of the Association and its General Assembly

- The Association shall have a General Assembly of its members and an elected Executive Committee.
- The General Assembly is the highest body of the Association.
- A General Assembly shall have a quorum when more than half of the members of the Association are present.
- The General Assembly shall be convened at a time and in a place that is decided upon by the Executive Committee.
- The General Assembly shall elect the members of the Executive Committee and shall adopt the bylaws, the work programme and activity report of the Association.
- The General Assembly shall consider and decide upon issues raised by any member of the Association.
- The General Assembly shall adopt decisions by a simple majority vote.

Composition and responsibilities of the Executive Committee

- The Executive Committee of the Association shall consist of a Chairperson, a Secretary, an Accountant and a Treasurer.
- The responsibilities of the Executive Committee shall be to:
  - Define the individual and collective shares of activities of the members and notify them accordingly;
  - Carry out any activity required to ensure that the bylaws are implemented;
  - Solicit means of capacity building for the local community;
  - Ensure that the property and resources of the Association are put into appropriate use;
  - Receive, examine and decide upon suggestions forwarded by members; and
  - Have the accounts of the Association audited by a qualified auditor.
- Each member of the Executive Committee shall serve for five years. However, the General Assembly can, if it so wishes, extend the term of office of any one of the members of the Executive Committee by another five years. Nevertheless, a person shall not be elected for the same office consecutively for more than two terms.
- The Chairperson shall legally represent the Association in all matters. He/She shall call and conduct the General Assembly as well as present the report of the activities of the Association. In the event that the Chairperson is unable to attend a General Assembly, he/she shall authorize one of the other members of the Executive Committee to stand in for him/her.
- Any bank account of the Association can be operated only through the joint signatures of the Chairperson and the Accountant.
- The Accountant shall keep all the accounts of the Association.

Appointing a third party of evaluators

○ We, the members of our Association, have agreed that the secretariat of our kebele in which our Association is found shall appoint a neutral third party of evaluators to examine the effectiveness of implementation of these bylaws at intervals and present its report to the Executive Committee of the Association.

○ Each member is obliged to provide any information concerning the Association required of him/her by the evaluators assigned by the kebele secretariat.

○ The evaluation report compiled by the kebele secretariat shall identify any weaknesses of the Association and suggest possible solutions. The report shall be presented by the Executive Committee to the General Assembly to be discussed and, as appropriate, for issues raised in it to be decided upon.

Offence and penalty

○ Any person who, without permission from the Chairperson, fails to attend a General Assembly has committed an offence and shall pay a penalty of .... Birr[2] for the first offence, ... Birr for the second offence and .... Birr for the third offence.

○ Any person whose animals are found in, or who drives another person's animals into, any area that has been closed off from free range grazing has committed an offence and shall pay to the Association a penalty of.................... Birr per animal for the first offence, ... Birr per animal for the second offence and .... Birr per animal for the third offence.

○ Any person whose animals are found in, or who drives another person's animals into, another member's holding which has been prohibited for free range grazing has committed an offence and shall pay to the Association a penalty of .... Birr per animal for the first offence, ...Birr per animal

---

2   Birr is the currency in Ethiopia. In other countries, it is the appropriate local currency that would be specified.

for the second offence and .... Birr per animal for the third offence.

- Any person who damages any dam or wall built to stop gullying, any terrace, any bund or any plants established for erosion control shall be compelled to repair the damage and, in addition, shall pay to the Association a penalty of .... Birr for the first offence, ... Birr for the second offence and .... Birr for the third offence.

- Any member who fails to protect or care for shrubs, trees or grasses planted around his/her parcel of land or any rehabilitating gully in his/her land has committed an offence and his/her case shall be presented to the General Assembly for decision taking.

- Any member who causes any problem in the implementation of a plan of the Association or who fails in his/her personal or share of collective responsibility in the implementation of an assigned activity shall have his/her case reviewed and decided upon by the General Assembly.

Date of coming into force of the bylaws

These bylaws have come into force as of .... , when the woreda (district) secretariat registered them as having been approved by the woreda administration following successive approvals by the General Assembly of our Association, the signing of each of our members, and the approval by the kebele administration.

Names and signatures of all members

Name	Signature
1	....
2	....
3	....
.	
x	....

ROLES OF STAKEHOLDERS

The main stakeholders include, but are not restricted to, the Environmental Protection Authority at the federal government level, the respective regional environmental protection office, the zonal (sub-regional) administration, the woreda administration and the kebele administration.

The possible roles that these stakeholders may play are detailed as follows. It should be emphasized that the roles described are meant to serve as examples, and the specific set of stakeholders and their respective roles are best defined by each local community and each woreda on a case-by-case basis.

The Environmental Protection Authority

The Environmental Protection Authority shall continue to prepare various documents that will help in the formulation and implementation of local community and woreda plans for environmental management for sustainable development and in the carrying out of related activities.

It can also carry out related supporting activities.

It can organize meetings to evaluate the manuals together with various experts drawn from the region in question.

It can organize the training of trainers drawn from among experts working in the concerned region and woreda or woredas.

It can provide technical support for, as well as monitor the implementation of, the woreda or local community plan for environmental management for sustainable development.

It can provide support for cross visiting programmes to exchange experiences.

Regional environmental protection office

The regional environmental protection office can create opportunities for the participation of bureaus of agriculture and rural development, other governmental and non-governmental organizations, educational and research institutions, the private sector etc., in the formulation and implementation of local community and woreda plans for environmental management for sustainable development.

It can organize cross visit programmes for exchange of experiences, especially within its region.

It can support the initiatives being taken by the regional, zonal (sub-regional) and woreda administrations.

It can monitor and give technical support for the implementation of the woreda and local community plans for environmental management for sustainable development.

It shall submit reports on the implementation of the woreda and local community plans for environmental management for sustainable development in the region to the Environmental Protection Authority.

It shall prepare its regional state of the environment report based on the reports it receives from woredas and submit it to the Environmental Protection Authority.

It shall carry out the necessary follow-up to ensure that local community associations for environmental management for sustainable development and their bylaws are given legal recognition.

In consultation with the Environmental Protection Authority, it shall monitor developments in environmental management for sustainable development and make suggestions for improvement and additional action.

Zonal (sub-regional) administration

The zonal administration can work together with the regional environmental protection office to support the implementation of the plans for environmental management for sustainable development of the woredas within its domain.

Woreda administration

The woreda administration shall convene meetings of all stakeholders in its domain to raise their awareness so that they can support and contribute to the implementation of its planned activities.

To this effect, it shall nominate an expert who shall be responsible for following up the implementation of local community plans for environmental management for sustainable development in its domain, it shall monitor the activities of the appointed expert and provide him/her the support needed to carry out his/her duties.

It shall convene a meeting or meetings of all the kebele administrations within its domain and explain to them the specifics of planning and implementation of local community and woreda plans for environmental management for sustainable development.

It shall collate the local community plans for environmental management for sustainable development in its domain into its own plan for a woreda-wide implementation and have the woreda-wide plan for environmental management for sustainable development approved by an assembly of the representatives of all the kebeles in its domain.

It shall compile its woreda state of the environment report on a yearly basis.

It shall encourage the diffusion, especially within its domain, of the successful experiences of its local communities through local cross visits to exchange experiences.

Kebele administration

It shall support the local communities in its domain to develop and implement their respective local community plans for environmental management for sustainable development.

It shall help its local communities obtain recognition from the woreda level for their respective plans for environmental management for sustainable development and especially for their bylaws for implementing these plans.

It should motivate other stakeholders within its domain to support the planning and implementation of local community plans for environmental management for sustainable development and shall prohibit activities, especially by outsiders, that could take place in violation of the respective local community plans and their respective implementing bylaws.

It shall help in the diffusion of successful experiences within and among local communities, especially within its domain.

It shall carry out other related activities required of it by its woreda administration.

CRITERIA FOR SELECTING PILOT WOREDAS FOR DEVELOPING WOREDA ENVIRONMENTAL MANAGEMENT FOR SUSTAINABLE DEVELOPMENT PLANS

The following criteria are merely being suggested. It is preferable that the environmental protection office of a given regional government develops the criteria that it finds appropriate for choosing woredas for initiating pilot plans for environmental management for sustainable development. This is likely to make the choice of criteria appropriate for the objective reality in each region.

The woreda should be one where:

- Land degradation is a serious environmental problem;
- Repeated droughts have been occurring, and desertification is a threat;
- Food shortage is a frequent problem;
- Activities by non-governmental organizations have been minimal;
- The local administration is keen to work towards sustainable development;
- The environmental and developmental conditions make the experiences gained there useful for other woredas in the region; and
- The monitoring of activities and associated developments by stakeholders will not incur much inconvenience.

SUCCESSES AND CHALLENGES IN ECOLOGICAL AGRICULTURE: EXPERIENCES FROM TIGRAY, ETHIOPIA

Sue Edwards,
Tewolde Berhan Gebre Egziabher
and Hailu Araya

CONTENTS

FIGURES

TABLES

ABOUT THE AUTHORS

Sue Edwards is the Director of the Institute for Sustainable Development (ISD), Addis Ababa, Ethiopia, and has been co-editor of the eight-volume "Flora of Ethiopia and Eritrea" since 1984. She is a taxonomic botanist, teacher and science editor by profession.

Tewolde Berhan Gebre Egziabher is the Director General of the Environmental Protection Authority of the Federal Democratic Republic of Ethiopia. He is a plant ecologist who has been a university academic staff member and an academic administrator as well as a negotiator on biodiversity issues.

Hailu Araya is the Sustainable Community Development team leader of the ISD. He is a geographer who joined ISD after completing his Masters in community resource management at Addis Ababa University in 2001.

ACKNOWLEDGEMENTS

The authors and the ISD would like to thank the Third World Network (TWN) and the Swedish Society for Nature Conservation (SSNC) for their continuing support for the work with the farmers and agricultural professionals in Tigray from 1996 (TWN) and 2005 (SSNC) until the present. They would also like to thank the Food and Agriculture Organization of the United Nations (FAO) for its interest and financial support to improve the data collection of crop yields, have them statistically analysed for publication, and for its invitation to present the findings in the International Conference on Organic Agriculture and Food Security, held 2–5 May 2007 in FAO, Rome, and now for this opportunity to produce an updated version of the 'Tigray Project' experience.

Also to be acknowledged is the continuous support and interest of the Bureau of Agriculture and Rural Development (BoARD) of Tigray, particularly the agricultural professionals working in the woredas in partnership with ISD, their administrations and the farmers, without whose efforts the increased crop productivity and improvements to their environments would not have taken place. It has been our pleasure to work closely with all of them.

EDITORIAL NOTES

The data from the Tigray Project were collected by the woreda agricultural professionals under the supervision of Arefayne Asmelash and Hailu Araya, entered for analysis by Lette Berhan Tesfa Michael, and analysed statistically by Drs Kindeya Gebre Hiwot and Fitsum Hagos of Mekelle University.

ISD is a member of the Third World Network.

INTRODUCTION
Overview of the project

The Tigray Region of Ethiopia is highly degraded, posing difficult challenges to farmers. The degraded environment contributes to low agricultural production, in turn exacerbating rural poverty.

This paper documents the results of an experiment in sustainable development and ecological land management in Tigray. The "Tigray Project", as it is often referred to, demonstrates that ecological agricultural practices such as composting, water and soil harvesting, and crop diversification to mirror the diversity of soil conditions can bring benefits to poor farmers, particularly to women-headed families. Among the benefits demonstrated are increased yields and productivity of crops, an improved hydrological cycle with raised water tables and permanent springs, improved soil fertility, rehabilitated degraded lands, increased incomes, increased biodiversity, and increased mitigation and adaptation to climate change.

The project is farmer-led, and builds on the local technologies and knowledge of the farming communities. Local communities have been empowered and they now develop legally-recognized bylaws to govern their land and other natural resources management activities.

The successes of the project have led to its expansion to include many more communities in the Tigray Region and in the rest of the country. This happened because the government has now adopted the approach used by the project as its main strategy for combating land degradation and for eradicating poverty from Ethiopia.

Background of the project

Land is degrading fast globally. According to the World Resources Institute (WRI) *et al.* (1998), the Earth's soil was in 1998 eroding at rates of between 16 and 300 times its rate of formation. Understandably, soil erosion is fastest in mountainous areas and Ethiopia is mountainous. But, this estimate shows that even people in areas that do not look prone to soil erosion should fight land degradation to stop it before it makes their land uninhabitable.

Such an attempt at reversing land degradation in Ethiopia, called "Sustainable Development and Ecological Land Management with Farming Communities in Tigray", is a broad-based open-ended experiment by farmers and local experts. The main aim of the project is to find out if a community-based ecological approach to rehabilitating the land and improving crop production through the application of ecological principles can both reverse land degradation and improve the livelihoods of poor smallholder farmers. The project motivates local communities to develop their own bylaws to apply ecological principles and to protect their other interests and thus improve the lives of their members. This approach has come to be referred to as the "Tigray Project" because it was started in willing communities in the Tigray Regional State in northern Ethiopia.

Administratively, Ethiopia is divided into nine regional states and two city administrations, below which are zones and then "woredas", the latter of which can be taken as equivalent to districts. Woredas are made up of parishes called "tabias" in Tigray and "kebeles" in other regions. Each tabia or kebele thus consists of several villages, though the villages are often not clearly delimited since the homesteads are usually scattered over the landscape.

The Tigray Project started in 1996 in four local communities in the central, eastern and southern parts of the Tigray Regional State. In 2009, the Institute for Sustainable Development (ISD) was working with the Bureau of Agriculture and Rural Development (BoARD), woreda experts and development agents to continue implementing the Tigray Project in 45 tabias in 12 woredas in the Tigray Region, in 33 kebeles in Meqdela Woreda in the Amhara Region, in two kebeles in Gimbichu Woreda in the Oromiya Region and in four kebeles in Arba Minch Zuria Woreda in the Southern Nations, Nationalities and Peoples Region. The experience from the Tigray Project is expanding fast in all the crop cultivating parts of Ethiopia. This has been made possible by the inclusion of the practices of the Tigray Project in the extension system of the country thanks to a push by both the Ministry of Agriculture and Rural Development and the Environmental Protection Authority (EPA).

The main activities of the Tigray Project are: training and follow-up on compost making and use including monitoring impacts on crop yields; water and soil conservation activities; restricting free range grazing and feeding animals from cut grass and branches of woody plants; making community ponds, small dams and river diversions to catch and hold water for use in the dry season; promoting and encouraging innovator farmers in water harvesting, bee keeping and use of biopesticides based on indigenous knowledge; supporting women-headed and elderly families (they are the poorest of the poor) through supplying seeds of spices and training in raising fruit and forage tree seedlings for sale to their neighbours; training unemployed girls who complete formal schooling to equip them with skills for earning an income; experience sharing through cross visits; and supporting the use of new and easy to manage technologies such as treadle pumps.

Challenges for Ethiopia

Ethiopia is a land-locked country in the north-east of Africa in what is often referred to as the "Horn of Africa", 32°42'–48°12' East longitude and 3°24'–14°53' North latitude with an area of 1.13 million km². Its topography is very diverse, encompassing mountains over 4 000 m above sea level, high plateaus around 2 000 m above sea level, deep and wide gorges cut by rivers and arid lowlands including the Dallol Depression in the north-east, which goes down to 110 m below sea level in the Afar Triangle within the Great African Rift Valley (Figure 1 and EMA, 1988).

The 2007 population and housing census showed the total population of Ethiopia to be 75 million, growing at 2.6 percent a year, of which about 84 percent is rural (FDRE, 2008). More than 49 percent of the population lives in the highlands above 2 200 m altitude, 11 percent lives below 1 400 m and 40 percent in between 1 400 and 2 200 m altitude. Human and animal pressure is higher in the highlands than in the lowlands (Assefa, 2003; EPA, 2003; FAO, 1986).

FIGURE 1

Map of Ethiopia showing the location of Tigray Region

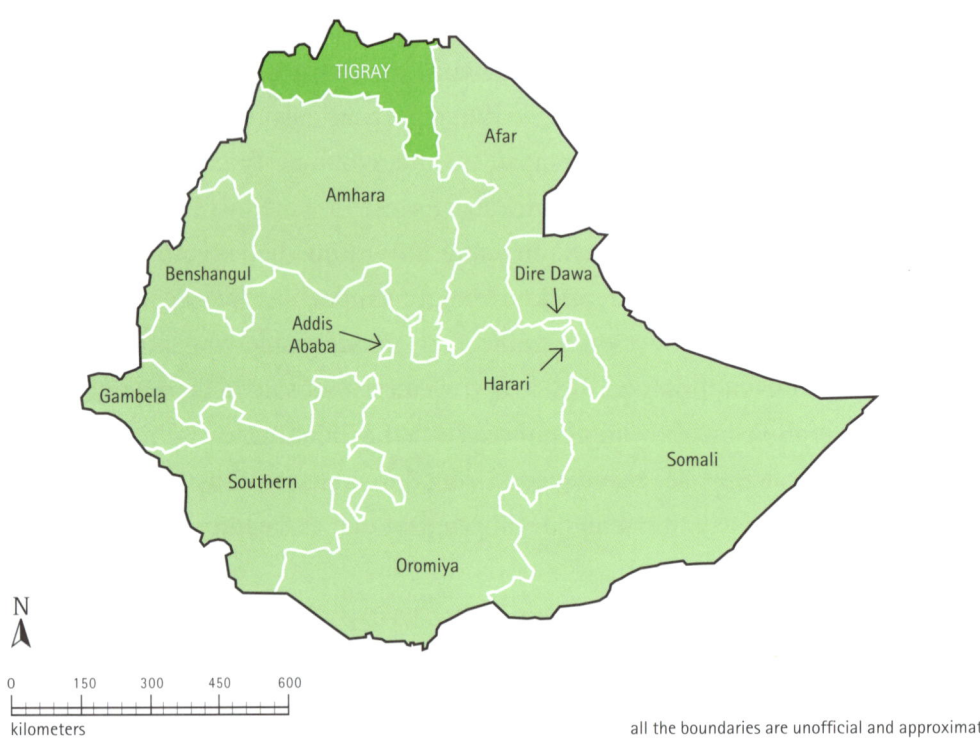

all the boundaries are unofficial and approximate

Source: FAO

The south-westerly summer winds are the most important of the country's three moisture-bearing wind systems (FAO, 1986). Originating from the Atlantic Ocean, they bring the greatest amount of moisture during the main rainy season (May/June–August/September). The small rains (February–April) originate from the Indian Ocean and bring rain to the central and eastern highland areas (EMA, 1988; Daniel, 1977). The mean annual rainfall is highest (above 2 700 mm) in the south-western highlands, gradually decreasing to 100 mm or less in the north-eastern lowlands. The mean monthly temperature varies from 45°C (April–September) in the Dallol Depression of the Afar lowlands to less than zero degrees at night in the highlands (November–February) (EPA, 2003; FAO, 1986).

The country currently faces a number of environmental challenges resulting directly or indirectly from human activities, exacerbated by rapid population growth and the consequent increase in the need to exploit the natural resources unsustainably (Asseffa, 2005). Estimates of the annual soil loss range from 400 t/ha in the cultivated highlands to 100 t/ha in the arable lowlands. The total removal is calculated as 1.5–1.9 billion tonnes of topsoil annually with an average total productivity loss from cropland of 1.8 percent (FAO, 1986). The misuse of natural resources caused by need includes burning animal dung as fuel instead of using it as a soil conditioner.

Crop cultivation in Ethiopia has a long history of at least 5 000 years (Clark, 1976), and implements for cutting and grinding seed have been found in stone age sites, such as Melka Konture by the Awash River in central Ethiopia, dating back much earlier. Just when crop cultivation started in Ethiopia has not been determined, but its long history is also reflected in the high agricultural biodiversity, including endemic crops, the best known of which is the cereal, teff (*Eragrostis tef*). The high diversity in crop species and genetic diversity within crops is a reflection of the environmental and cultural diversity of Ethiopia (Engels and Hawkes, 1991).

Many crops that are known to have their centres of origin in the fertile crescent of south-west Asia, for example durum wheat (*Triticum durum*), now have their highest genetic diversity in Ethiopia. The treatment of *Triticum* for the "Flora of Ethiopia and Eritrea" recognizes a highly variable endemic species, *T. aethiopicum*, which is more usually considered as a subspecies or variety of *T. durum* (Phillips, 1995). Other important crops with high genetic diversity in Ethiopia include the cereals – barley (*Hordeum vulgare*), finger millet (*Eleusine coracana*) and sorghum (*Sorghum bicolour*); the pulses – faba bean (*Vicia faba*), field pea (*Pisum sativum* including the endemic var. *abyssinicum*), chick pea (*Cicer arietinum*), lentil (*Lens culinaris*), fenugreek (*Trigonella foenum-graecum*) and grass pea (*Lathyrus sativus*);

the oil crops – linseed (*Linum sativum*), niger seed (*Guizotia abyssinca*), safflower (*Carthamus tinctorius*) and sesame (*Sesamum indicum*); and the root crops – enset (*Ensete ventricosum*), anchote (*Coccinia abyssinica*), "Oromo or Wollaita dinich" (*Plectranthus edulis*), and yams (*Dioscorea* spp.). Over 100 plant species used as crops in Ethiopia have been identified (Edwards, 1991).

European travellers, e.g. Alvares at the beginning of the sixteenth century (Alvares, 1961) and later ones, describe the productivity and health of the highland agriculture – crops, domestic animals and people – and compare this with the depressed situation in much of Europe at that time. Poncet (1967), who visited Ethiopia between 1698 and 1701, described his experience with the words, "no country whatever better peopled nor more fertile than Aethiopia". He described even the mountains he saw as all well cultivated "but all very delightful and covered with trees".

However, since 1974, Ethiopia has been portrayed as a food deficit country with its people and animals suffering from drought and famine. In 2002 around 14 million people (20 percent of the population) required aid as food because of the failure of the rains in much of the eastern parts of the country. In 2008 and early 2009, there was again drought throughout the eastern, south-eastern and southern parts of Ethiopia as well as much of Kenya because of the failure of the rains.

Starting in the second half of the nineteenth century, efforts to build an administratively centralized Ethiopian state as a reaction to European colonialism in other parts of Africa systematically destroyed local community governance because it was suspected that such communities could become possible allies of colonialists. Loss of local governance undermined local natural resource management with loss of protection of woody vegetation, lack of repair of old terraces, and general undermining of any attempts at communal management of natural resources. The feudal landlord system was maintained with the bulk of the population existing as serfs. As Ethiopia entered into the world market, these landlords mined the resources of the peasants with nothing going back to the land. Civil war exacerbated these impacts. The most visible physical impacts have been gully formation eating away the soil with vegetation recovery prevented by free range grazing and the unregulated felling of trees for firewood and other purposes.

There were no inputs in technologies or ideas to help these peasant farmers improve their productivity. They had to continue to rely for their survival on their indigenous knowledge and the rich agricultural biodiversity that they had developed, but were unable to continue effectively developing and using collectively for fear of political reprisal.

Then, in 1974, Emperor Haile Selassie and the feudal system of control over farmers and their land were removed in a revolution that organized the whole population into local, nominally self-governing, organizations with their own elected officials. Under the military government, called the "Derg", there were massive efforts at land rehabilitation through mass mobilization for soil and water conservation, planting of tree seedlings, and the provision of external inputs through cooperatives. However, administration remained centralized and coercive, and overall productivity did not increase. The farmers continued to be ordered about and exploited as had been done under the over-centralized feudal regime. There were also frequent and disruptive redistributions of land. The farmers had no possibility for taking collective decisions on natural resources management and no interest or incentives to invest in improving the land.

In 1991, the military government was overthrown. A new constitution that required decentralization of power and encouraged local community governance was adopted in 1995. In 1993, the Sasakawa-Global 2000 Project was launched to provide high external inputs – principally chemical fertilizer – to farmers.

Although the reasons for poverty are thus complex, those behind the poverty of rural Ethiopia have mainly arisen from total neglect by previous governments, with the land thus being mined to feed an emerging urban population and for trade. These pressures and deliberate interference have led to the collapse of traditional land management systems for maintaining environmental integrity in general and soil fertility in particular. Due to such problems, the total area under food crop production, mainly cereals, has increased while grain production per unit area has remained generally low, less than one tonne a hectare (Assefa, 2003).

The 2004 Human Development Report (UNDP, 2004) classified Ethiopia as one of the least developed countries in the world. Agriculture accounted for more than 75 percent of total exports, over 85 percent of employment, and about 45 percent of the gross domestic product (GDP). Coffee alone made up more than 87 percent of the total agricultural exports. Hides and skins were the next most important export items as raw, processed or manufactured goods (EPA, 2003).

The Government's Sustainable Development and Poverty Reduction Program has identified agriculture as the key sector in which to devote efforts for accelerating socio-economic development and reducing poverty (MoFED, 2002). However, the interventions for agriculture must be ecologically, economically and socially sustainable if the people are to find a way out of poverty that supports improved human welfare and basic rights.

Conditions in Tigray

Tigray Region is found in northern Ethiopia and is generally regarded as the most degraded part of the country. Most of the region is highland, but the eastern part includes the escarpment facing the Great East African Rift Valley. The national population census of 2007 estimated the population of Tigray as 4.3 million with an annual growth rate of 2.5 percent occupying an area of just over 50 thousand square kilometres (FDRE, 2008). The average population density of the region is 80 persons/km^2, with high concentrations in the Eastern, Southern and Central Zones where it is 131, 122 and 115 persons/km^2 respectively (Figure 2 and CSA, 2002).

Average annual rainfall in Tigray is 800–1 000 mm in the west and the highlands of the south dropping to 400 mm in the extreme east. In most parts, it averages between 400 and 600 mm/year (EMA, 1988). The precipitation occurs mostly during a short summer (end of June to mid-September) rainy season, often falling as intense storms (FAO, 1986; Hunting, 1976). Except in some remote areas and around churches, by 1975 the natural dryland forest and woodland vegetation of the Tigray Region had been destroyed. This was because of overgrazing, the progressive increase in demand for fuel wood and land for cultivation. Serious soil erosion and yield reduction have occurred for at least the last 100 years (Hunting, 1976). However, since the 1980s, many areas of natural forest and woodland have reappeared on hillsides following agreements by local communities to restrict access by people and grazing animals to these areas. A study using photographs from Tigray by researchers from the University of Ghent in Belgium and Mekelle University show that much of Tigray now has a better vegetation cover than in 1975, and even than that found in 1868 (Nyssen *et al.*, 2008; Nyssen *et al.*, 2009). However, the recent development of settlements to accommodate landless families, as well as the rise in the urban population continues to put serious pressure on the forest and woodland resources of the region (Hailu and Hailemariam, 2006, unpublished).

FIGURE 2

Population density of Tigray Region by woreda

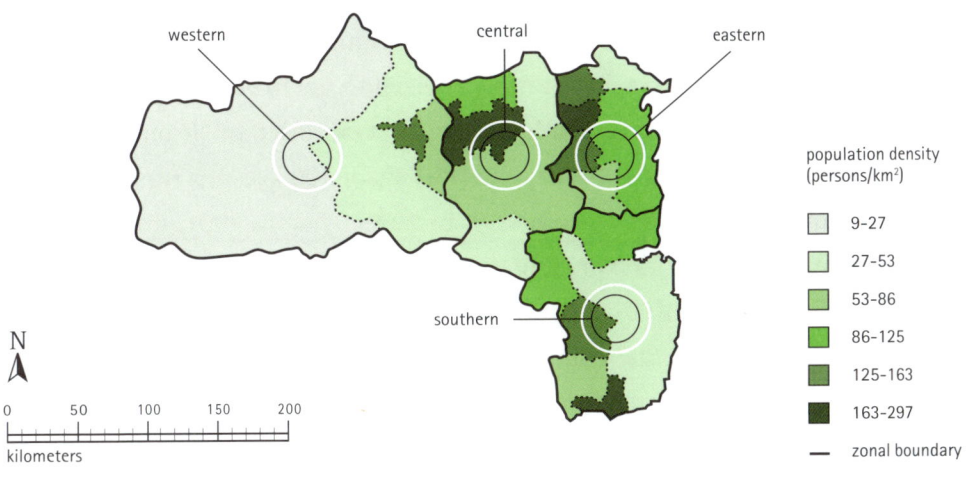

Problems of chemical inputs

Using chemical fertilizer is recent in Ethiopia. It started in the late 1960s along with the launching of integrated agricultural programmes and projects (EPA, 2003). For millennia, farmers have been using traditional systems of fallowing, crop rotations, manure and wood ash to maintain soil fertility and their crop yields. The Sasakawa-Global 2000 (SG 2000) programme started in 1993 and, in 1995, it was incorporated into the National Extension Intervention Programme of the Ministry of Agriculture. The aim was to boost food crop production through a focused campaign to get farmers to use chemical fertilizer along with high yielding varieties (HYVs), when available, and pesticides. However, the programme mainly gave a strong support to the adoption of diammonium phosphate (DAP) and urea fertilizers through credit schemes and subsidized prices. This was largely because, owing to the diverse nature of the environment, most of the crops had had no HYVs bred, and even those that had been bred, were suited only for limited areas.

Since 1998, the subsidy on chemical fertilizer has been withdrawn while the price has risen. Despite that, by 2001, only around 5 percent of the smallholder farmers of the country, particularly those growing maize and wheat, had become habituated to using chemical fertilizer. But that year, the price dropped out of the bottom of the maize market and the retail price fell to the lowest it had been since 1994/95, well below the cost of production. The retail price in Nekemt (Western Ethiopia) in May 2001 was ETB 31 (around USD 4) for 100 kg. This was 41 percent below the market price in the same month in 1995 (Alemu, 2001, unpublished). At the same time, the farm gate price of maize in western Ethiopia as a whole was reported to have fallen to ETB 18 (USD 2.2) for 100 kg.

In 2002, many parts of the country were hit by drought with the result that yields were very low and some crops failed completely; the government requested food aid for nearly 20 percent of the total population. The long-term problem is the large number of rural families facing chronic food insecurity who cannot recover even in a good growing season. Due to the crop failure owing to the drought, many farmers were heavily in debt, and during the growing season of 2003, they withdrew from taking fertilizer especially in areas where moisture stress is frequent. The farmers in these areas say that chemical fertilizer "burns" crops in the field. It also causes soil salinization, damages soil structure, and reduces the number and diversity of smaller organisms living in the soil.

Owing to the unavailability of HYVs, the extension programme allows farmers to select and use the best of their own local varieties. Smallholder farmers use their own saved seed for over 90 percent of the total sowing in the country (Hailu, 2003, unpublished). However, by 2004, the Ethiopian Seed Agency was promoting and distributing seed of higher yielding crop varieties to be grown either for local food processing or for export, and rapid erosion of farmers' varieties was taking place in some of the moister and more accessible areas.

Pesticides are little used except for dealing with the migratory pests, particularly armyworm, desert locust, quelea bird and localized insect swarms, such as Pachnoda beetles on sorghum and the endemic Wello Bush Cricket on cereals. The biggest users of agrochemicals are the few large-scale government-operated farms, particularly those growing cotton. However, smallholder horticultural production from small-scale irrigation systems is expanding and the farmers are making increasing use of chemical inputs. The continuing introduction of HYVs is also attracting pests and chemical control is being made available to farmers. The use of pesticides is often with little or no understanding of either how to store them safely, or how to mix and use them correctly. The misuse of pesticides and fertilizers is damaging human health with severe impacts and even deaths recorded, as well as polluting the surrounding environment (Tadesse and Asferachew, 2008).

Nonetheless, an experiment in sustainable development and ecological land management conducted with farmers in Tigray demonstrates that ecologically sound agricultural practices can succeed and bring benefits to farmers, without recourse to harmful and expensive chemicals (Tewolde Berhan et al., 2004).

THE TIGRAY PROJECT
Climate change and policy environment for the project

The Environmental Policy of the Federal Democratic Republic of Ethiopia had not been approved by the government when the Tigray Project started in 1996, but it was used by the Tigray Project as the basis for identifying technologies that farmers could use. The Environmental Policy emanates from the Constitution of the Federal Democratic Republic of Ethiopia, viz: "All persons have the right to a clean and healthy environment." (Constitution of the Federal Democratic Republic of Ethiopia, Paragraph 1 of Article 44)

The Environmental Policy of the Federal Democratic Republic of Ethiopia was approved in 1997 (FDRE, 1989 EC/1997 GC). Climate change had then not pre-occupied the whole world as it does now, though it was very evident to those involved in examining the global environment. The United Nations Framework Convention on Climate Change (UNFCCC), which was opened for signature in the Rio de Janeiro Conference on Environment and Development in 1992, was already in force in Ethiopia, but the Kyoto Protocol was only just coming into existence. Therefore, the Environmental Policy of Ethiopia does not directly refer to the Kyoto Protocol; neither does it use the now familiar terms of "climate mitigation" and "climate adaptation". However, it deals with the substance of both these terms.

Understandably, the reduction in polluting the atmosphere that the Environmental Policy deals with is limited to attempts at political influence to appeal to the major emitters of greenhouse gases to mend their ways. Article 3.9 (b) states: "To recognize that even at an insignificant level of contribution to atmospheric greenhouse gases, a firm and visible commitment to the principle of containing climate change is essential and to take the appropriate control measures for a moral position from which to deal with the rest of the world in a struggle to bring about its containment by those countries which produce large quantities of greenhouse gases."

Ethiopia's greenhouse gas emissions are limited almost entirely to the relatively few vehicles and aeroplanes that the country runs. The generation of electricity is almost entirely from hydropower. Virtually all household energy is derived from biomass. The firewood and cow dung that are burnt become new firewood and new grass and hence new cow dung during the subsequent rainy season.

Climate adaptation in Ethiopia is, of course, an Ethiopian affair, and the Environmental Policy deals with it in detail. In fact, virtually the whole of the Environmental Policy is directly or indirectly on adaptation to, and simultaneous mitigation of, climate change. For example, Article 3.9 (a) states: "To promote a climate monitoring programme as the country is highly sensitive to climatic variability." Environmental information is treated in more detail in Article 4.7 with its seven sub-articles, and environmental research in Article 4.8 also with its seven sub-articles. The Environmental Policy is found in the website of the EPA at <www.epa.gov.et>.

Ethiopia's commitment to tackling climate change is shown in Article 3.9 (c), which states: "To recognize that Ethiopia's environmental and long-term economic interests and its energy prospect coincide with the need to minimize atmospheric inputs of greenhouse gases as it has a large potential for harnessing hydro-, geothermal and solar energy, none of which produce pollutant gases in significant amounts, and to develop its energy sector accordingly." Article 3.9 (e) states: "To recognize that the continued use of biomass for energy production makes no net contribution to atmospheric pollution as long as at least equal amounts of biomass are produced annually to compensate this and to maximize the standing biomass in the country through a combination of reforestation, agroforestry, the rehabilitation of degraded areas, a general revegetation of the land and the control of free range grazing in the highlands, and to seek financial support for this from industrialized countries for offsetting their carbon dioxide emissions." Article 3.5, with its nine sub-articles, deals with renewable energy in much more detail.

Ethiopia's economy is dominated by the agriculture sector. Even in the future when Ethiopia will have developed strong industrial and service sectors, agriculture will always remain important. Ethiopia has the potential to always continue being suitable for agricultural production to supply the needs of people and other animals.

To adapt agriculture to climate change, and even to continue in a world without any climate change, there will always be a need for good soils, forests and other vegetation cover, genetic resources and water resources. Article 3.1, with its 19 sub-articles, deals with the care that everyone should use to sustainably manage the soils of the country in the unpredictability of floods, droughts and winds that climate change is set to exacerbate. Article 3.2, with its nine sub-articles, covers the management of forests, woodlands and other woody biomass resources so as to maximize wood production for both climate change mitigation and adaptation. Article 3.3, with its 11 sub-articles, shows both the existence in Ethiopia of genetic resources that are of global importance, as well as the management that these genetic resources need for agricultural systems that would be robust now and in the coming future of a changing climate.

Article 3.4, with its ten sub-articles, deals with Ethiopia's plentiful water resources unevenly distributed both geographically and temporally. The Ethiopian mountainous landscape imposes an inherent obstacle to using water resources effectively within the country. It is, on the whole, from the higher and more arid parts along the east that water flows to the lower and more moist parts along the west – i.e. from where it is needed more, to where it is needed less. This paradoxical situation can be alleviated through effective water harvesting and management upstream without compromising Ethiopia's responsibility to maintain the flow of water coming out of its mountainous highlands to its lower-lying neighbouring countries.

The management and use of local resources, including soil, water, vegetation, and agricultural and other forms of biodiversity, can become effective only when the local communities that live in and use each locality are each well informed and effectively organized. Paragraph 4 of Article 50 of the Constitution emphasizes this fact. It states: " ... Adequate power shall be granted to the lowest units of government to enable the People to participate directly in the administration of such units." Article 4.10 of the Environmental Policy, with its nine sub-articles, aims to make environmental information available to local communities. Paragraph (d) of Article 4.5 aims to ensure gender equality in access to environmental information and in empowerment for environmental management. Article 4.2, with its seven sub-articles, aims to empower local communities of men and women that enjoy equal rights and equal access to information on environmental management so that they can organize themselves on equal terms and effectively care for and effectively use their soil, water, vegetation, and biodiversity resources. They would then remain robust and maintain a robust environment to face the vagaries of a changing climate.

How the principles set out in the Environmental Policy are being applied is described in the section on "Scaling up the Tigray Project".

Origin and development of the project

"Is there sufficient biomass to make adequate quantities of compost?" This is the question most often raised whenever there is any suggestion that Ethiopia could use ecological principles to increase crop yield.

In 1995, Dr Tewolde Berhan Gebre Egziabher, then leading the project to develop a Conservation Strategy and Environmental Policy for Ethiopia, was asked by some government officials to design a project that could be promoted among farmers living in degraded areas in order to improve the productivity of their land and rehabilitate their environments while at the same time contributing to carbon sequestration and adapting agriculture to climate change. Some officials in the Tigray Region expressed an interest in the project and a workshop was held in 1995 to launch it and identify local stakeholders.

The project was called "Sustainable Development and Ecological Land Management with Farming Communities in Tigray", though it is usually referred to as 'The Tigray Project'. The local authorities call it the "Sustainable Agriculture/Development Project". The ISD was established to implement it. Activities started in 1996 in partnership with the BoARD and Mekelle University. The direct beneficiaries of the project have been the local farming communities, their local development agents, experts and administrations.

From the beginning, the project has been funded by the Third World Network (TWN), an international NGO network with its head office in Penang, Malaysia. In 2006, TWN published the experiences of the Tigray Project (Hailu and Edwards, 2006), which included preliminary findings from the impacts of the use of compost on yields of crops in farmers' fields.

As from 2005, the Swedish Society for Nature Conservation (SSNC) has also provided funding to ISD for its work in Tigray through supporting the development of a nursery to supply farmers with seedlings of fruit trees and to continue its work with farming communities, local agricultural professionals and their administrations. It has also supported publications, including a poster on making compost to supplement the compost manual in Tigrinya (the local language) published in 2002 (Arefayne, 1994 EC/2002 GC). In 2007, SSNC invited ISD to participate in a study of ecological agriculture based on improving local ecosystem services to identify general patterns in agro-ecosystems that improve production based on local resources and services. The objective of the study is to identify and analyse examples from local agricultural communities in four continents that will be examined with systems ecological and resilience theory as a starting point. The first study visit was to ISD's work with the farmers, farming communities and local agricultural professionals in Tigray. It took place in October–November 2007. The outcome of this study was published in 2008 with the title "Ecological in Ethiopia: Farming with Nature Increases Profitability and Decreases Vulnerability" (SSNC, 2008).

In 2006, the FAO Natural Resources Department provided funding to help ISD compile existing and collect additional yield data to produce a database of over 900 crop yield records from farmers' fields, and pay for entry into a computer and statistical analysis of the data (Edwards *et al*, 2007). The results of this analysis are presented in the next section.

The first four communities to become involved in the project in 1996 were Adi Nifas in the densely populated Central Zone west of the ancient town of Axum, Adi Abo Mossa in the Southern Zone on the sloping land beside Lake Hashenge, and Gu'emse and Ziban Sas in the Eastern Zone, generally regarded as the driest and most degraded part of the region. The work started with an experienced senior extension officer of the BoARD, Arefayne Asmelash, consulting the communities about the problems with their environment, and discussing possible solutions, none of which required high external inputs but demanded commitment from the community members and local agricultural professionals.

The suggested technologies (basket of choices) were designed to build on the respective local community's own traditional systems of farming and land management with some selected additions from field-tested traditional and scientific knowledge gleaned from within Ethiopia and other countries.

The farmers were thus stimulated to select from a 'basket of choices' made up from suggestions put forward by ISD, BoARD and the farmers themselves. The key components of the project were as follows:

- Making and using compost to restore soil fertility, to sequester carbon and to avoid getting into a debt trap by buying chemical fertilizer on credit;
- Restricting free range grazing by domestic animals and encouraging cutting and carrying of fodder to feed them, in order for the natural vegetation to recover and biodiversity to be protected;
- Digging trench bunds for catching both water and soil along field boundaries;
- Halting and rehabilitating gullies through building check dams at regular intervals;
- Terracing slopes both to minimize soil erosion and to maximize rain water infiltration;
- Making ponds to store rain water for animals and for making compost in the dry season;
- Planting small multipurpose trees – particularly *Sesbania sesban* – local grasses and legumes on the bunds; and
- Formulating and using bylaws to coordinate the activities of the members of the local community and thus to control access to and use of local biological resources including the restrictions of free range grazing by domestic animals.

Each of the four original communities chose a different entry point into the project.

Adi Nifas had two gullies eating away the farmers' fields and the farmers agreed to try the other components in the project if the gullies could be stopped. Check dams were built by the community at intervals in the gullies, which successfully arrested further development of the gullies; Adi Nifas became the model community demonstrating how all the components reinforced each other to improve the lifestyles of the farmers. By 2001, a spring and small stream that had dried up for decades reappeared and farmers below the project site started a small irrigation scheme because of the water flow that enabled them to harvest at least two crops a year.

Ziban Sas had had its grazing area of 13.5 ha destroyed by sheet erosion, and the water table had dropped to below 12 m. Therefore, the farmers, particularly the women, focused first on rehabilitating their grazing land, and then became interested in the other components. With grazing animals kept off the land, the local grasses and other herbs soon reappeared, and check dams and trench bunds led to better infiltration of water. But of more significance for the women was the raising of the water table so that they were able to use their local well again.

Gu'emse also had a very large gully created by powerful flooding in the rains from a large catchment area. This was eating away the farmers' fields and bringing deposits of silt onto their land. Attempts to control this gully were not successful until 2005 when appropriate materials and expertise were made available to the community. The initial failure to check the gully somewhat discouraged the farmers although a core of about 45 households continued making and using compost and also initiated making a communal pond to catch the flood water. Reclaiming the large gully required inputs of cement and wire-mesh, locally called "gabion", to hold the stones in place and the success has reversed the community's attitude to the project. The protection of their achievements features prominently in their bylaws and they have planted many trees and grasses to help stabilize both the catchment area and the gully.

Adi Abo Mossa is a relatively well-endowed area with good rainfall and a large grazing area around the lake. The farmers had been using increasing amounts of chemical fertilizer. Therefore, the implementers of the project wanted to encourage the farmers to adopt composting in order to reduce the threat of eutrophication in Lake Hashenge. The farmers make very good quality compost in large quantities and most have been able to give up using chemical fertilizer. But this is the only component of the project adopted by these farmers. This is because neither increasing biomass through restricting free range grazing, nor raising the water table are important to them.

In all four communities, the farmers quickly saw in one or two growing seasons the impact of compost on crop yields and the improved water holding capacity of the soil in their fields. In May 1998, representatives of the farmers and their development agents from the four communities made cross visits to each others' areas ending up with a two-day discussion and general evaluation of what they had seen of the project's activities. This encouraged all the communities, except those in Adi Abo Mossa, to take up all the components of the project.

These activities have increased the biomass in the farms and their surroundings. Therefore, policy makers have come to realize that the answer to the question "Is there sufficient biomass to make adequate compost?" is "If farmers want, they can make enough compost, especially at the end of the growing season".

In 2001, the Ethiopian government issued its Policy on Rural Development Strategies and Guidelines (MoI, 2001) and set up a supra-ministry, the Ministry of Agriculture and Rural Development, to coordinate activities for its implementation. The policy regards environmental rehabilitation as an essential precondition for increasing productivity. It emphasizes the need to improve local marketing infrastructure, and also to develop more agricultural products so as to diversify the economic base of the country.

The Environmental Policy of Ethiopia has incorporated a basic principle similar to one adopted in organic agriculture: "To ensure that essential ecological processes and life support systems are sustained, biological diversity is preserved and renewable natural resources are used in such a way that their regenerative and productive capabilities are maintained, and, where possible, enhanced...; where this capacity is already impaired to seek through appropriate interventions a restoration of that capability" (FDRE, 1989 EC/1997 GC). This policy supports the development of more specific policy and regulations for organic agriculture; the government has already issued an "Ethiopian Organic Agriculture System" (FDRE, 2006).

There is an increasing awareness of the importance of producing healthy fruits and vegetables organically for the expanding educated middle-class and expatriate market in Addis Ababa. For example, Genesis Farm started in 2001 and its organic production now covers over 40 hectares. The farm combines dairy and poultry production together with growing vegetables, fruits and ornamental plants. It sells certified products on the export market. However, there is also a fast expanding local market and it is interesting to note that none of the items sold by Genesis are more expensive than other locally produced items, and several are even cheaper. Local workers are able to buy their vegetables from the farm shop. There are now several other farms and urban agriculture associations within and in the peri-urban area of Addis Ababa that are supplying ecologically produced healthy fruits and vegetables to local shops and restaurants.

The international trade in organic products is an expanding niche market that Ethiopia is geographically well situated to exploit. Communities in the southern and south-western parts of the country have formed cooperative societies and unions growing and exporting Arabica coffee with an organic and fair trade label. This has enabled farmers to obtain prices similar to those in 1996, when the international coffee prices started to collapse. A survey in 2008 found that Ethiopia had 137 822 ha of land certified as organic being used by 110 861 farmers mostly organized in 40 companies and cooperative unions, with a few in individual farms. There were four foreign, but no national, certifying organizations working in Ethiopia.

In 2005, the Ministry of Agriculture and Rural Development produced a "Guideline on Community-based Participatory Watershed Development" together with an Annex. These are comprehensive documents incorporating the experience of many practitioners. Participation is seen as central to the planning, implementation, monitoring and evaluation of watershed development. The documents include information cards on the technologies that can be used, including one for compost making. Although ISD is not mentioned specifically, its experience in Tigray has helped in shaping the development of this document.

In 2006, environmental issues were incorporated into the revised second phase of Ethiopia's Poverty Reduction Strategy Programme (PRSP). The PRSP is taken as the guideline for all development support by the donor community in Ethiopia. One of the biggest programmes that has been launched by the government is the "Productive Safety-Net Programme", which is channelling funds directly to woredas for use in public works activities as well as in environmental rehabilitation focusing on watersheds.

The impact of compost on crop yields

The first exercise in getting yield data from the use of compost was in 1998. The data showed that using compost gave similar yield increases as the use of chemical fertilizer. In November 2001, these results were presented at a workshop in Axum including representatives from all the regions as well as key government offices in Addis Ababa. The data were published as an annex to the booklet, "Natural Fertilizer" (Edwards, 2003). ISD was urged by the senior staff of the BoARD to continue monitoring the impact of compost on crop yields, and this has been done each year since 2000 in some of the local communities with which ISD has been working.

The method used to obtain the yield data is based on the crop sampling system of the FAO. The fields for taking the yield samples are selected together with the farmers and are chosen to represent the most widely grown crops. Three one-metre square plots are harvested from the field. These plots are placed in the field to reflect the range of conditions of the crop, i.e. both well grown and not so well grown areas are selected from the centre and towards the edge of the field. The harvested crop is then threshed and the grain and straw are weighed separately. The plot data are recorded along with the name of the farmer, the crop and the treatment as well as the location and the date. The farmer keeps the straw and grain. The straw is important because it is the main animal feed, especially during the dry season.

The average from several fields of the same crop in the same area given the same treatment is calculated. "Check" means the field received no treatment, although it may have received compost in one or more of the previous years. "Compost" is for fields treated with mature compost. The rates of compost application range from around 3.5 t/ha in poorly endowed areas, such as the dry Eastern Zone of the region (Ziban Sas and Gu'emse), to around 15 t/ha in the moister Southern Zone (Adi Abo Mossa). "Chemical fertilizer" is for fields treated with diammonium phosphate (DAP) and urea. The recommended rates are 100 kg/ha of DAP, and 50 kg/ha of urea.

The yields are converted to kg/ha for comparison. It should, however, be remembered that farmers' fields are very small, often less than a quarter of a hectare, and that the total cultivated area of most farmers is less than one hectare.

An important feature of the Tigray Project is that it is to a large extent led by the farmers. They choose which crops to treat with compost and which with chemical fertilizer – some also compare the use of manure, while a few combine compost and chemical fertilizer.

Crop yields in 2002

The year 2002 had poor rainfall in most parts of Tigray and only a few farmers got grain from their fields. However, it was a very good demonstration of the effect of compost on the moisture holding capacity of the soil. The drought was very severe, even in western Tigray, but a farmer in Adi Aw'ala applied compost to his field of sorghum and got a grain yield equivalent to 2 t/ha and a straw yield of 3 t/ha; an untreated field gave only the equivalent of 0.8 t/ha of grain and 1.5 t/ha of straw. He also grew maize with compost doubling the yield of the check (see Figure 4).

Figures 3–6 give the yields for barley, maize, teff and wheat from 14 communities. Only four of the communities, Adi Abo Mossa, Adi Gua'edad, Adi Nifas and Gergera, had yields from three of the four crops. Teff is the fastest maturing of the crops, and yields were obtained from nine of the communities, the best being 3.9 t/ha from a chemically treated crop in Enda Maino, where the compost treatment gave 2.9 t/ha. But, overall, the fields that had received compost gave higher yields than those treated with chemical fertilizers.

It is interesting to see that the yields from the checks in Adi Abo Mossa and Adi Nifas were close to those from the fields that had received compost or chemical fertilizers. These two communities had been making and using compost since 1996, and its residual effect in the soil is thus clearly seen. The very low yields for all crops in Adi Nifas in 2002 are the result of the drought. More normal yields are given in Table 2.

FIGURE 3

Barley yields in six communities in 2002 (kg/ha)

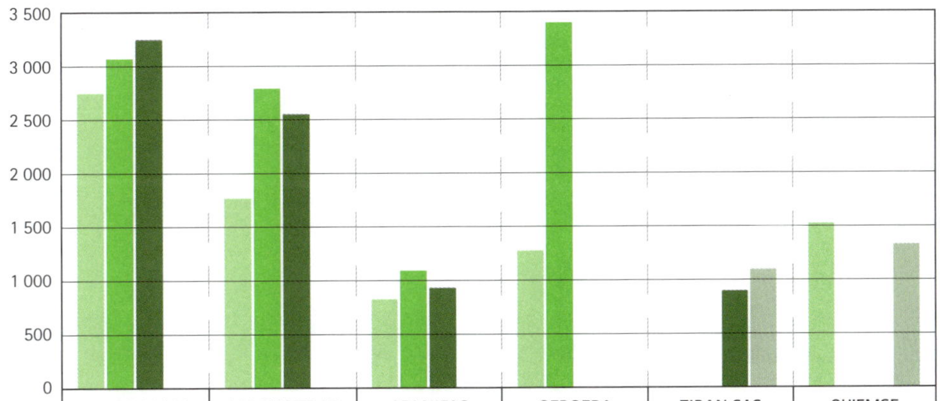

	ADI ABO MOSSA	ADI GUA'EDAD	ADI NIFAS	GERGERA	ZIBAN SAS	GU'EMSE
check	2 765	1 782	833	1 283	–	1 535
compost	3 090	2 808	1 100	3 420	–	–
chemical fertilizer	3 263	2 562	933	–	900	–
manure	–	–	–	–	1 100	1 333

FIGURE 4

Maize yields in five communities in 2002 (kg/ha)

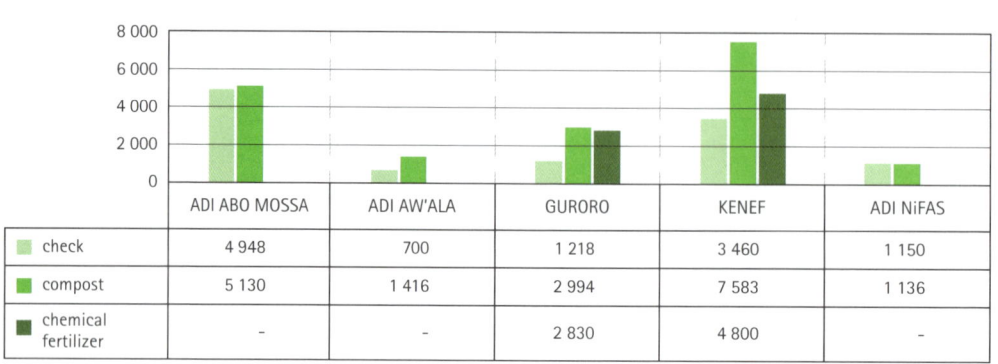

	ADI ABO MOSSA	ADI AW'ALA	GURORO	KENEF	ADI NiFAS
check	4 948	700	1 218	3 460	1 150
compost	5 130	1 416	2 994	7 583	1 136
chemical fertilizer	–	–	2 830	4 800	–

FIGURE 5

Teff yields in nine communities in 2002 (kg/ha)

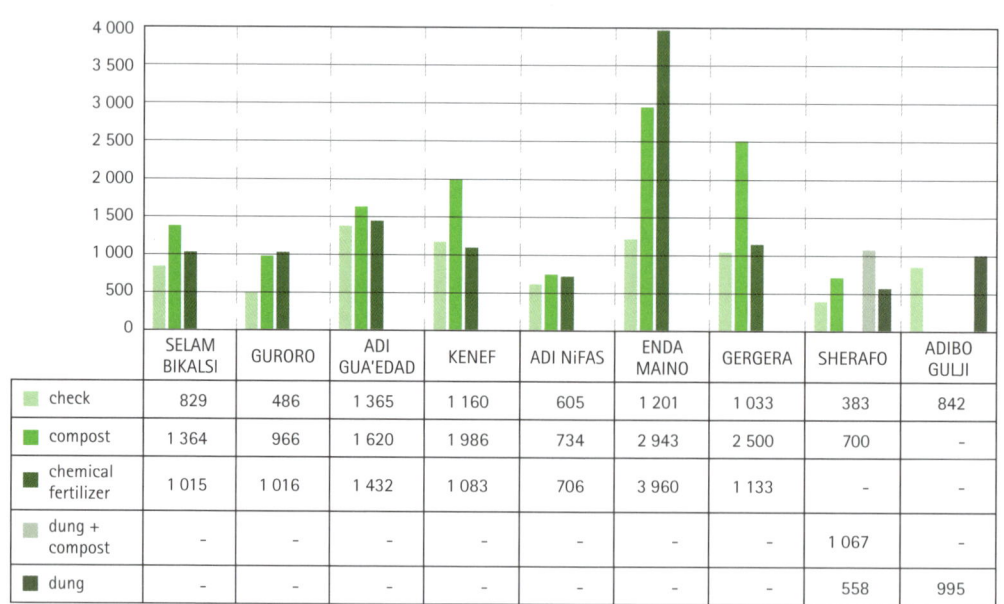

	SELAM BIKALSI	GURORO	ADI GUA'EDAD	KENEF	ADI NiFAS	ENDA MAINO	GERGERA	SHERAFO	ADIBO GULJI
check	829	486	1 365	1 160	605	1 201	1 033	383	842
compost	1 364	966	1 620	1 986	734	2 943	2 500	700	–
chemical fertilizer	1 015	1 016	1 432	1 083	706	3 960	1 133	–	–
dung + compost	–	–	–	–	–	–	–	1 067	–
dung	–	–	–	–	–	–	–	558	995

FIGURE 6

Wheat yields in six communities in 2002 (kg/ha)

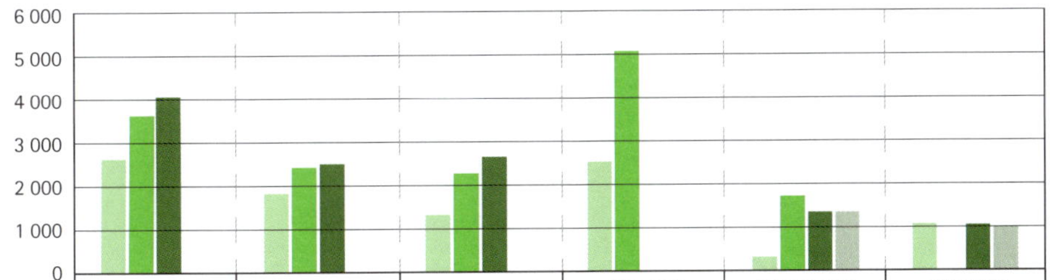

	ADI ABO MOSSA	ADI GUA'EDAD	ENDA MAINO	GERGERA	ZIBAN SAS	GU'EMSE
check	2 783	1 882	1 333	2 716	400	1 113
compost	3 820	2 402	2 376	5 083	1 800	–
chemical fertilizer	4 030	2 522	2 783	–	1 450	1 110
dung	–	–	–	–	1 500	925

Crop yields in 2003/4

The rainy season was better in 2003 and yields were generally higher than in 2002. In spite of the differences on them of the impact of the failure of the rains in 2002, as can be seen from Figures 3 and 5, the communities of Adi Nifas and Adi Gua'edad are geographically close to each other and it was considered useful to compare the use of compost and chemical fertilizer by the two communities, and also to give an indication of the value of the crops as well as the expense of using chemical fertilizer for the farmers. In 2003, 100 kg of DAP was ETB 270, and 50 kg of urea was ETB 107. Thus the total cost for applying these to one hectare in 2003 was ETB 377 as cash or ETB 433.55 through credit including 15 percent interest.

Average grain prices in ETB per 100 kg in the local market in 2003 in Tigray were 225 for teff, 155 for maize, 200 for faba bean, 170 for barley, 170 for wheat, and 145 for finger millet. The averages were calculated from prices collected from different markets in the region.

It can be argued that the opportunity cost for making compost should also be included. However, it should be noted that compost is made when there is little need for other farm activities, i.e. during the gap between crops flowering and maturing for harvest, and the opportunities for off-farm work are not high at this time. All the materials for making compost are freely available locally. Therefore, the net income does reflect what the farmer can expect to get from his/her crops.

The Adi Nifas farmers had been using compost for seven years while those of Adi Gua'edad had only used compost for two years. The yields are shown in Tables 1 and 2.

TABLE 1

Grain yields in kg/ha, expenses and returns in ETB* for Adi Nifas (2003)

CROP	INPUT	YIELD	GROSS INCOME	FERTILIZER COST	NET INCOME
Faba bean	Compost	4 391	13 173	–	13 173
	Check	2 287	6 861	–	6 861
Finger millet	Compost	2 650	4 505	–	4 505
	Check	833	1 416	–	1 416
Maize	Compost	5 480	8 768	–	8 768
	Check	708	1 133	–	1 133
Teff	Compost	1 384	3 875	–	3 875
	Chemical fertilizer	1 033	2 892	377	2 515
	Check	739	2 069	–	2 069
Wheat	Compost	2 250	5 625	–	5 625
	Chemical fertilizer	1 480	3 700	377	3 323
	Check	842	2 105	–	2 105
Barley	Compost	1 633	3 266	–	3 266
	Check	859	1 718	–	1 718

*In 2003, 10 ETB = 1 Euro, or 8.5 ETB = 1 USD

In Adi Nifas only one farmer used chemical fertilizer for his teff, and the yield was less than what he got from his field with compost. But in Adi Gua'edad, all the farmers used chemical fertilizer, even on their faba bean. However, the yield of the faba bean was less than half of the yield from using compost, and the economic return from applying the compost was much higher.

TABLE 2

Grain yields in kg/ha, expenses and returns in ETB* for Adi Gua'edad (2003)

CROP	INPUT	YIELD	GROSS INCOME	FERTILIZER COST	NET INCOME
Faba bean	Compost	2 900	8 700	–	8 700
	Chemical fertilizer	1 100	3 300	377	2 923
	Check	766	2 298	–	2 298
Finger millet	Compost	2 000	3 400	–	3 400
	Chemical fertilizer	1 433	2 436	377	2 059
	Check	500	850	–	850
Maize	Compost	2 000	3 200	–	3 200
	Chemical fertilizer	1 133	1 813	377	1 436
	Check	680	1 088	–	1 088
Barley	Compost	2 193	4 386	–	4 386
	Chemical fertilizer	1 283	2 566	377	2 189
	Check	900	1 800	–	1 800
Wheat	Compost	1 020	2 550	–	2 550
	Chemical fertilizer	1 617	4 043	377	3 666
	Check	590	1 475	–	1 475
Teff	Compost	1 650	4 620	–	4 620
	Chemical fertilizer	1 150	3 220	377	2 843
	Check	390	1 092	–	1 092

*In 2003, 10 ETB = 1 Euro, or 8.5 ETB = 1 USD

These results from the two communities are a reflection of what has happened in all the areas where the farmers have learnt how to make and apply compost. In the first year or two, they continue to use chemical fertilizer, but as they gain confidence in making compost and also see its residual effects in restoring soil fertility, they stop buying chemical fertilizer. It is interesting to note that after farmers stop using chemical fertilizer, no reduction in yield has been recorded as they 'convert' to using only compost. This is probably because the use of chemical fertilizer is relatively recent, i.e. only since 1995 or even later. More importantly, this is probably because they do not plant the so-called high yielding varieties, which have a high demand for chemical fertilizer. Instead, they plant their best yielding traditional varieties, which have been bred to thrive in organically fertilized fields.

Statistical analysis of crop yields sampled between 2000 and 2006

Starting from 2000, yields have been taken from plots in farmers' fields in 19 communities in eight of the 53 woredas of Tigray Region (Table 3). The majority of the communities (17) are found in the drought prone areas: Alamata of the Southern (two communities), and all parts of the Eastern (six communities) and Central (nine communities) Zones. The soils of these areas are generally prone to land degradation and the rainfall is erratic. However, two communities are found in better endowed areas: Adi Abo Mossa on the land sloping down to the shores of Lake Hashenge of Southern Tigray, where the soils are deep, rainfall more reliable and some farmers have larger cultivated areas and large herds of cattle, and Adi Aw'ala in Western Tigray where the rainy season is generally two to four weeks longer than in the rest of the region.

Between 2000 and 2006, grain and straw yield data were taken separately from 974 plots. The names of the 11 crops from which observations were recorded are given in Table 4. But four of these were dropped from the final statistical analysis because each had less than ten observations. This left seven cereal and two pulse crops in the final statistical analysis.

TABLE 3

List of local communities from which crop yield data were taken between 2000 and 2006

ZONE	WOREDA	TABIA	COMMUNITY
Southern Tigray	Ofla	Hashenge	Adi Abo Mossa (0)
	Alamata	Lemat Seelam Beqalsei	Adi Abo Golgi Seelam Beqalsei
Eastern Tigray	Sa'esi'e Tsada Amba	Sendeda Mai Megelta Agamat	Tsebela Zeban Sas (0) Gu'emse (0)
	Kilte Awla'elo	Mai Weyni	Sherafo
	Atsbi-Wonberta	Hayelom	Gergera Enda Maino
Central Tigray	Tahtai Maichew	Mai Berazio	Adi Nifas (0)
		Akab Se'at	Adi Gua'edad
		Ruba Shewit	Adeke Haftu
		Mai Siye	Mai Tsa'ida
		Kewanit	Hagere Selam
		Adi Guara	Tselielo
		Adi Hutsa	Kenef
	Kolla Tembien	Guroro Miwtsa'e Worki	Shimarwa Adi Reiso
Western Tigray	Tahitay Adyabo	Adi Aw'ala	Adi Aw'ala
Total	8	18	19

Key – (0) refers to communities where work started in 1996/7, the others joined the project later.

The data were analysed using the statistical program, STATA. The average grain and straw yields converted from g/plot to kg/ha for each treatment for the nine crops are given in Table 5. The table also gives the number of observations included in the analysis for each crop and treatment. The average of the grain and straw yields for each treatment for the seven cereal crops is shown in Figure 7.

The data for the nine crops were subjected to linear regression analysis by treatment based on the values obtained from fields where compost was applied, chemical fertilizer (DAP and urea) was applied and no input (check) was applied. The null hypothesis used was that the treatments have no impact on the yields.

TABLE 4

List of crops from which yield data were recorded, 2000–2006

	CROP	SCIENTIFIC NAME	REMARKS
1	Barley	Hordeum vulgare	Many farmers' varieties are grown
2	Durum wheat	Triticum durum	The most widely grown wheat
3	Finger millet	Eleusine coracana	Not grown as widely as in the past
4	Hanfets	Hordeum vulgare + Triticum durum	A mixture of barley and durum wheat grown in areas prone to erratic rainfall and generally poor soils
5	Maize	Zea mays	Grown more for the fresh cobs than the grain
6	Millet	Eleusine coracana	The same as finger millet – less than ten observations were recorded under this name
7	Sorghum	Sorghum bicolor	Grown more widely in the western lowlands than the highlands
8	Teff	Eragrostis tef	Ethiopia's endemic cereal with many varieties
9	Chick pea	Cicer arietinum	Not very widely grown – less than ten observation were recorded
10	Faba bean	Vicia faba	The most widely grown pulse, also known as horse bean
11	Field pea	Pisum sativum	More often grown mixed with faba bean than by itself
12	Haricot bean	Phaseolus vulgaris	A recent introduction by the BoARD – less than ten observation were recorded
13	Horse bean	Vicia faba	The same as faba bean – less than ten observations were recorded under this name

The probability that this null hypothesis could explain the results was found to be less than 0.05. In other words, the confidence limit was found to be above 95 percent. The increase in grain yields in fields where chemical fertilizer was applied was significantly higher (95 percent confidence limit) than in the fields where no input (check) was applied, and the increase in the grain yields in fields where compost was applied was also significantly higher (95 percent confidence limit) than in the fields where chemical fertilizer was applied. The significance in the differences among the straw yields for each treatment was similar. The differences among treatments in the yields of each of the crops were also similarly significant.

TABLE 5

Average yields by treatment in kg/ha for nine crops in Tigray, 2000–2006

CROP TYPE	AVERAGE YIELD (KG/HA)					
	CHECK		COMPOST		FERTILIZER	
	Grain	Straw	Grain	Straw	Grain	Straw
Barley	1 115	2 478	2 349	4 456	1 861	3 739
	(n=56)	(n=52)	(n=57)	(n=55)	(n=36)	(n=35)
Durum wheat	1 228	2 342	2 494	3 823	1 692	3 413
	(n=73)	(n=67)	(n=61)	(n=57)	(n=48)	(n=45)
Finger millet	1 142	2 242	2 652	4 748	1 848	3 839
	(n=16)	(n=16)	(n=14)	(n=13)	(n=8)	(n=7)
Hanfets	858	2 235	1 341	3 396	1 199	2 237
	(n=31)	(n=31)	(n=31)	(n=31)	(n=29)	(n=29)
Maize	1 760	3 531	3 748	4 957	2 900	3 858
	(n=31)	(n=20)	(n=41)	(n=31)	(n=25)	(n=13)
Sorghum	1 338	2 446	2 497	3 662	2 480	4 433
	(n=14)	(n=13)	(n=11)	(n=10)	(n=5)	(n=5)
Teff	1 151	2 471	2 143	3 801	1 683	3 515
	(n=106)	(n=94)	(n=75)	(n=66)	(n=71)	(n=68)
Faba bean	1 378	2 121	2 857	4 158	2 696	3 783
	(n=20)	(n=17)	(n=23)	(n=24)	(n=3)	(n=3)
Field pea	1 527	1 201	1 964	1 625	0	0
	(n=9)	(n=9)	(n=9)	(n=9)		

'Hanfets' is a mixture of barley and durum wheat

n = Number of records for each treatment and crop

The use of compost also gave higher yields than the use of chemical fertilizer, though differences in the yields from compost and from chemical fertilizer were not as great as the differences between the use of compost and the check. For sorghum and faba bean, the yields from the use of compost and chemical fertilizer were similar. But the yield difference for all the other crops was greater with that from the compost treatment being always higher than that from the use of chemical fertilizer.

FIGURE 7

*Average grain and straw yields (kg/ha) for seven cereal crops, based on
the averages for each crop, Tigray, 2000-2006*

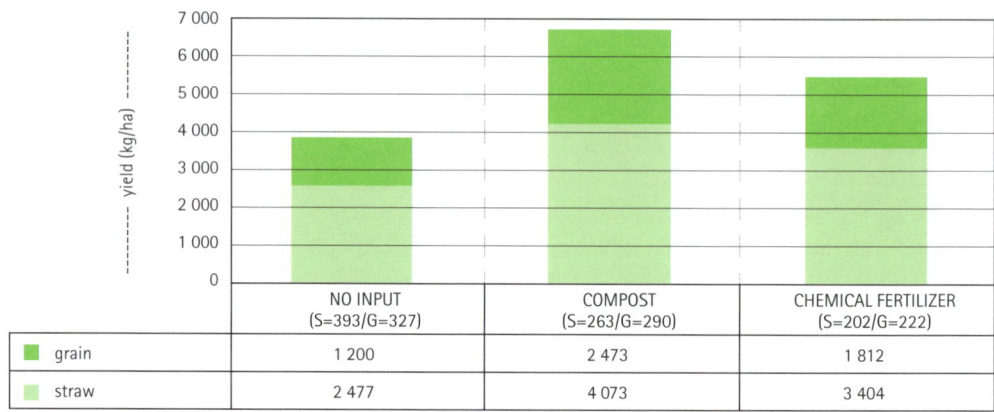

	NO INPUT (S=393/G=327)	COMPOST (S=263/G=290)	CHEMICAL FERTILIZER (S=202/G=222)
■ grain	1 200	2 473	1 812
■ straw	2 477	4 073	3 404

s=number of observations for straw yield
g=number of observations for grain yield

The proportion, expressed in percentages, of the grain in the total harvested biomass (grain + straw) for each of the nine crops is given in Table 6. For the cereal crops, the percentages of the grain in the harvest are given in Figure 8. The data are only indicative because the farmers usually leave long stubble up to 20 cm tall from their cereal crops in the field for domestic animals to graze on. For faba bean and field pea, however, all the above ground biomass is harvested. The results show that compost not only increases the overall biomass yield, but also increases the proportion of the grain to straw in the yield. The most striking crop is field pea where the proportion of grain in the total yield exceeded 50 percent for both the check and the compost treatment, but the field pea 'check' was probably grown in fields that had received compost in previous years – see Figure 9 and the discussion on it. For all the other crops, the proportion of grain in the total harvested yield ranged from 28 percent for hanfets to 35 percent for sorghum in check fields,

TABLE 6

Total biomass and percentage grain by crop in Tigray, 2000–2006 inclusive

| CROP TYPE | % GRAIN IN TOTAL BIOMASS YIELD (KG/HA) | | | | | |
| | CHECK | | COMPOST | | FERTILIZER | |
	% grain	total	% grain	total	% grain	total
Barley	31	3 593	35	6 805	33	5 600
Durum wheat	34	3 570	39	6 317	33	5 105
Finger millet	34	3 384	36	7 400	32	5 687
Hanfets	28	3 093	28	4 737	35	3 436
Maize	33	5 291	43	8 705	43	6 758
Sorghum	35	3 784	41	6 159	36	6 913
Teff	32	3 622	36	5 944	32	5 198
Faba bean	39	3 499	41	7 015	42	6 479
Field pea	56	2 728	55	3 589	0	0

"Hanfets" is a mixture of barley and durum wheat

from 28 percent for hanfets to 43 percent for maize in fields treated with compost, and from 32 percent for finger millet and teff to 43 percent for maize in fields where chemical fertilizer had been applied.

In 1998, when the first set of data were collected from plots in the four original communities, the grain yields of the cereals from the fields without any inputs (checks) were all, except for maize, below one tonne per hectare: 395–920 kg/ha for barley, 465–750 kg/ha for durum wheat, 760 kg/ha for finger millet, 590–630 kg/ha for hanfets, and 480–790 kg/ha for teff (Annex in Edwards, 2003). In the seven-year data set, only hanfets had an average grain yield for the check below one tonne per hectare (858 kg/ha). The average check yields for all the other cereals ranged from 1 115 kg/ha for barley to 1 760 kg/ha for maize. By 2006, the four original communities had been making and using

FIGURE 8

Average grain expressed as percentage, in grain plus straw yields for seven cereal crops, Tigray, 2000–2006

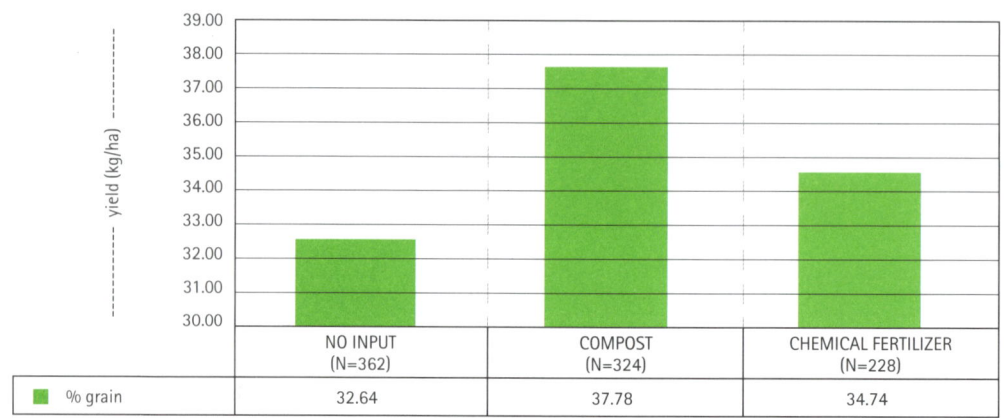

	NO INPUT (N=362)	COMPOST (N=324)	CHEMICAL FERTILIZER (N=228)
% grain	32.64	37.78	34.74

n=number of observations

FIGURE 9

Yields (kg/ha) for faba bean and field pea from Adi Abo Mossa, 1998 and 2002

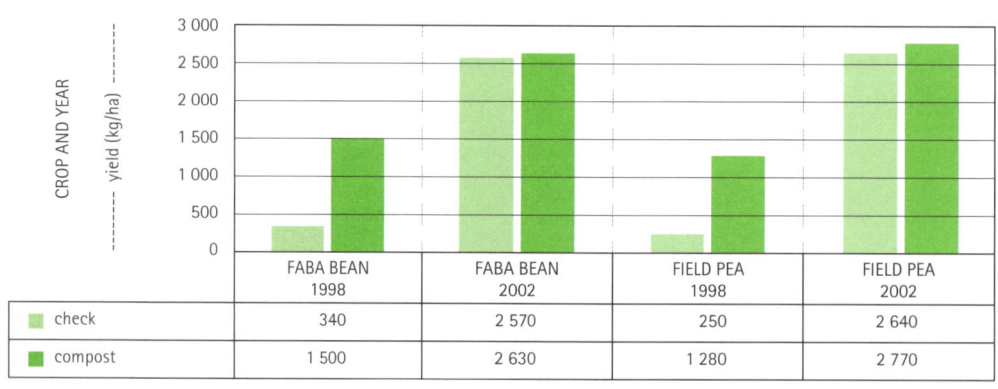

	FABA BEAN 1998	FABA BEAN 2002	FIELD PEA 1998	FIELD PEA 2002
check	340	2 570	250	2 640
compost	1 500	2 630	1 280	2 770

compost for ten years, all the others had been using compost for three to five years, and the higher average check yields were probably due to the residual effect of the use of compost in previous years.

The impact of compost on restoring soil fertility is well illustrated by data for grain yields of the pulses, faba bean and field pea, shown in Figure 9 for Adi Abo Mossa. The difference between the yields for the check fields and fields that had received compost was very large in 1998, but in 2002 there was hardly any difference – for both crops and both treatments, the grain yields were over two tonnes a hectare (Tewolde Berhan *et al.*, 2004). This similarity in yields is also seen for field pea in the seven-year data set in Table 5.

Farmers rarely use chemical fertilizer for legume crops but these results show that compost can increase yields much above untreated (check) fields. The increased yields of leguminous crops from the use of compost are important for the farmers as these crops have high market values (see Tables 1 and 2). They are also important components of the diet since very little meat is normally consumed by smallholder farming households.

The residual effect of compost in maintaining soil fertility for two or more years was soon observed and appreciated by the farmers. They are thus able to rotate the application of compost on their cultivated land and do not have to make enough to apply to all their cultivated land each year.

The reduction of difficult weeds, such as Ethiopian wild oats, *Avena vaviloviana*, and improved resistance to pests, such as teff shoot fly, have also been noted by the farmers. These impacts from the use of compost, including better resistance to crop diseases, have also been found by farmers practicing organic agriculture in France (Chaboussou, 1985).

One reason that compost has been able to significantly increase yields could be the fact that the farmers are still using their own varieties (also referred to as landraces), which have been selected by them in an organic environment where overall soil fertility is more important than just the amounts of the two major nutrients, nitrogen and phosphorus, supplied by urea and DAP. Dr Stephen Jones (personal communication) of the Washington State University and his colleagues have been breeding wheat for organic agriculture and they find that varieties that give high yields under organic conditions are different from those that give high yields with chemical fertilizer inputs. The farmers of Adi Abo Mossa apply the highest rates of compost to their fields. By 2004, they were complaining of lodging in their fields that had received compost. There also seemed to be a levelling off in the increase in yields indicating the need for the farmers to be involved in participatory plant breeding in order to select varieties that could respond with higher yields under an organic production system.

Economic and social impacts of using compost

There have been many positive impacts on the lifestyles of farmers who have taken to adopting ecological farm management to give them sufficient biomass for making and using compost. Because they do not have the financial burden of taking chemical fertilizer on credit, several farmers, including women-headed families, have improved their houses, bought additional animals such as chickens and milk cows, and beehives.

Many farmers have diversified their crop production. One interesting innovation now widely taken up by other farmers is intercropping. A farmer in Adi Nifas who intercropped tomato and chilli pepper in his teff field set the example. Scattering other crops such as cress (*Lepidium sativum*) and oil crops into teff fields is a traditional practice. These crops continue growing up after the rains finish and do not compete with the teff much. This is also the case with the tomato and chilli pepper. There is a high demand for these vegetables, particularly at the end of the rainy season before the irrigated vegetable areas have started to produce them. The farmer who started the practice claimed to get more from selling his vegetables than from the teff. Several other farmers in the Adi Nifas and Adi Gua'edad areas have now adopted this and other innovative combinations of intercropping with teff.

One farmer in Ofla had to share-crop his land because he did not have plough oxen. He heard about the use of compost and convinced his share-cropping partner to use it. When the harvest was sold, the farmer got sufficient income to buy one ox. The next year, he was able to buy a second ox and give up share-cropping.

A very poor widow with five children was unable to send them to school or even join in social functions because none of the family members had basic socially acceptable clothing and she had nothing to contribute to the communal meals. She then grew the spice, coriander, instead of a food crop. This is a high value crop. After just one harvest, she got sufficient income to send her children to school because she could buy them the clothing that they needed and she could participate in social functions because she could buy the clothing and food that she needed for herself.

The special needs of women

Tigray has a high number of women-headed families because the long civil war that ended in 1991 was the most intense there. Much effort has been made in the project to include households headed by women because these are generally among the poorest of the poor in their villages. Traditionally, women should not handle plough oxen and they therefore have to wait to have their land ploughed by a male relative or a neighbour. This puts the single woman at a double disadvantage. Her fields are planted later, and each day's delay in sowing reduces the final yield. Then, at harvest time, the woman has to share part of the harvest with the farmer who ploughs her field. Such households never have enough food and are often unable to join in social activities or send their children to school because of lack of adequate clothing. A few women have taken up ploughing their own fields, but this has generally been at a considerable social cost with ridicule and ostracism levelled at them by their neighbours and relatives. They have even had their animals poisoned or mutilated. ISD and the local officials try to make sure that the poor women in the communities are included in all activities and their achievements are recognized.

In 2003, ISD suggested that the long season crops, finger millet, sorghum and maize, could be raised as seedlings in nursery beds and then planted out when the main rains started. This has been readily adopted by women, particularly for finger millet, as they can plant seedlings, rather than broadcast seed, when their fields are ploughed and they can thus overcome the disadvantage of late planting. The first woman to try transplanting got the equivalent of over 7.2 tonnes/ha from the part of her field where she had applied compost and transplanted finger millet seedlings. Direct sowing into the other part of her field where she had also applied compost gave the equivalent of 2.8 tonnes/ha. The number of farmers, both women and men, who transplant finger millet has steadily increased each year with most of them getting good yields.

In 1998, ISD decided to find out if supplying women with seed of spices, which are high value crops, could help poor women break out of their poverty trap. This has been the case and there is now a high demand from women each year for seed of spices. These women grow and sell the spices and get sufficient income to buy food and clothing for their families. This approach to supporting poor women has been taken up by other non-governmental organizations (NGOs) working in Tigray.

Many women find it difficult to prepare compost, mainly because they cannot dig the pits. ISD sometimes provides financial support for labour to dig these pits. However, during compost workshops, poor women-headed households are deliberately chosen to have pits dug and filled by the workshop participants, including themselves, so that they can get a boost to start making and using compost.

The other main constraint to making compost is the need for water. Collecting water for her household can take a woman around three hours for each trip. Often she has to make this trip twice a day. It is, therefore, not possible for a woman to collect water also for making compost or for growing fruits and vegetables. But she can use waste water from her household activities. Many local innovators have explored ways to dig shallow wells of up to 12 metres deep (but usually less than 5 metres deep) and have developed innovative ways of lifting the water out of these wells. ISD has provided support for these innovators to dig shallow wells for poor women-headed households, and then encouraged the women to grow fruits and vegetables. Again this has helped such women and their families get out of the poverty trap. This form of support has now been taken up by the BoARD and other NGOs working in Tigray.

The community of Adi Abo Mossa used another approach to compost making that ensured that women could also be included. Groups of around ten households make their compost in large communal pits. This makes it possible for the women to be included because they help with providing composting materials, and then with taking the compost to the fields. But this is the only community with such a pooled compost making system.

In 2005, the Tahtai Maichew Woreda Administrator asked for poor women-headed families to be provided with small animals (sheep or goats). Elsewhere in the region, this has been found to be one way of helping women get an income to improve their livelihoods. In all, 13 women (six from the town and seven from the villages) were each given one male and four female goats or sheep. This was on the basis of a revolving fund as each woman would return one female animal to be given to other poor women or men. Although almost all the women got at least one new lamb, the women in the town faced a problem of finding sufficient feed for their animals in the town. One woman said that she was planning to move back to her village with her sheep as she could then find feed more easily in her village area. In 2006, this approach was expanded to three other woredas through the Nile Basin Initiative: Transboundary Micro-grants Program. Granting of the sheep or goats was tied to the women agreeing to control their movements and feed them through a cut and carry system.

In 2004, leaders in Tahtai Maichew Woreda selected 30 girls and young women who had completed tenth grade for skill training for income generation. The training continued for two and half months with the graduation ceremony in January 2006. Of the 30 girls, half were trained in traditional food preparation and the other half in basketry and some embroidery. The standard of workership was very high in both areas, and some of the food items that had been made were quickly sold out to the local people who had come for the graduation ceremony. The head of the Tahtai Maichew Woreda women's affairs office assisted the young women to form a cooperative and apply for micro-credit to develop their own businesses. Training in traditional food processing was scaled up to other woredas through a grant to ISD from the Nile Basin Initiative Transboundary Micro-grants Program.

The challenges

The communities that have taken up the Tigray Project are neighbourhood groups of between 15 and 90 households. Each community belongs to a tabia (sub-district) of 300 to over 500 households. The tabia, with its "baito" is the lowest level of official administration. A "baito" is the tabia's council elected by the people and is responsible for the administrative and socio-economic functions within its jurisdiction. Thus, the baito plays a decisive role in local governance. Local collective decisions are codified in bylaws formulated by the members of a local community convened by its elected leadership. There may be more than one such local community in a tabia. Legal affairs are adjudicated by a social court of the tabia, though a local community does also enforce its own bylaws.

The woreda is the next higher level of administration. Each woreda has its own administration consisting of an administrator and a cabinet of five experts and advisors. All the main sectors of the government are represented in the woreda by experts. Woreda-level coordinators direct the activities of development agents, who are in day-to-day contact with the farmers. It has been very important from the beginning that the woreda officials, experts and development agents understood and supported the aims of the project for its effective implementation.

At the level of the farmers, the improvements in livelihoods from the Tigray Project have been readily appreciated. However, these improvements, particularly the management of the natural resources in both the cultivated and uncultivated parts of a catchment as a whole, are bringing out challenges because a small community can develop and apply its own local bylaw only to its own members, but in order to resolve conflicts and problems with neighbouring communities, the local bylaw needs recognition and support from both the baito of the tabia, as well as from the woreda.

Local community level bylaws have established local committees to oversee the implementation of the project. These committees meet regularly to discuss and make decisions for local management. They have also set up a system of penalties for farmers who trespass their bylaws; for example, by letting their animals graze in protected cultivated and non-cultivated areas or cutting grass when they have not been authorized to do so.

Restricting grazing

The most challenging aspect of the project to implement has been restricting the grazing of cattle, sheep and goats.

Traditionally, only plough oxen, milking cows and young stock get special feeding treatment; all other domestic animals have to find feed for themselves. The animals are turned out in the morning and are guided and guarded by children during the day, including being driven to communal drinking places. In most parts of Tigray, there has been a drastic reduction in the areas of common grazing land as the demands for cultivated fields have increased with the increasing numbers of farming households. Traditionally, after the harvest is collected, the animals range freely over fields and patches of natural vegetation, while crop residues are kept on tops of houses and small trees or in specially fenced parts of the compounds for feeding special animals. This system of free range grazing has helped drive the deterioration of the environment as the animals trample more than they eat, destroy seedlings of woody plants and break down physical structures constructed for soil and water conservation.

It was, therefore, seen as essential that animal grazing had to be restricted if attempts to allow hillsides to be re-vegetated and increase biomass were to succeed. In fact, in the 1980s many of the villages in Tigray had already instituted a system whereby communities identified hillside areas as enclosures keeping out grazing animals as well as people and the native trees and grasses re-established. But, up to 2005, neither the region nor the woredas had developed a system or systems for the sustained use of the biomass of the enclosures. Hence farmers changing from free range grazing to cutting and carrying for their animals have not always found it easy to protect their recovering vegetation from free range grazing animals from neighbouring villages. With more communities adopting the restriction of animal grazing and with all levels of government throughout the country recognizing community bylaws, this problem is on the way to being solved.

One unexpected positive aspect of changing the grazing management has been the opportunity for children to go to school. ISD has heard that children who are required to stay out of school to herd animals resent this and it can cause division in families. Changing grazing management for families without children, such as elderly couples, is also a major problem as they do not have the time or energy to cut and carry feed for their animals.

The problem of restricted grazing was raised in a meeting in Tahtai Maichew in August 2005 and again in a compost training workshop in September 2005. The first was a meeting of around 40 farmers who had expressed their dislike of the Tigray Project. Most of these farmers were better off and had larger herds of animals than their poorer neighbours. They had only seen the project in terms of restrictions and had not realized the positive environmental impacts of controlling grazing on improvements to the general productivity of their areas, and particularly to the improvements to the local hydrology. After a long and intense discussion lasting half a day, the farmers agreed that they would try and cooperate with the aims of the project and control the movements of their animals.

The discussion in the compost training workshop in September 2005 emphasized the importance of having cooperation among farmers. The farmers stated their awareness of the importance of controlling the grazing of animals, but, they pointed out that their neighbours do not restrict the movements of their animals, and the local authorities, the baito committees and social courts, do not uphold the bylaws of the communities and help in penalizing the farmers that break the grazing rules. Therefore, they were not all prepared to control the grazing of their animals unless this could be enforced throughout their communities, i.e. by the baito.

However, the advantages of restricting grazing and using cut and carry is gradually being seen by many farmers. All the levels of government, supporting the BoARD, are working to convince farmers to restrict the free range grazing of their animals, and the regional government has adopted a policy called 'zero grazing'. This has brought a new challenge – the need for improved and easily available forage on the bunds between the fields as well as in the rehabilitated gullies. *Sesbania* and tree lucerne (*Chamaecytisus proliferus*) are popular, some farmers have also introduced pigeon pea (*Cajanus cajan*), while elephant grass is being widely planted. In many communities, starting from Adi Nifas, the farmers have introduced local species of tall thatching grasses (*Hyparrhenia* spp.) on the bunds. But the demands for seed and planting material are much higher than the ability of local institutions to make them available. It is also becoming clear that research and development is needed to further intensify the economic productivity of the whole local ecosystem with emphasis on the areas dependent solely on rainfall.

Forests

Traditionally, local communities usually have some form of common law (unwritten understandings and systems) for using their natural resources, particularly water resources and trees. But these common laws are not written and it has sometimes been difficult to have them recognized by the formal codified (written) legal system of the country; a hangover from the history of the centralization of state power of the nineteenth and early twentieth centuries. Until the development of the system of enclosures by local communities, the best protected areas of natural vegetation were the small patches of sacred forests around churches and monasteries, in Muslim burial grounds and around holy springs. These areas have been protected by the strong beliefs of local people, not through official guarding.

The other forest/woody areas are "community forests", mostly of eucalyptus planted during the military government of the 1970s or early 1980s. After the change of government in 1991, local communities throughout the region have designated hillside enclosures where indigenous tree and shrub species, particularly acacias and some juniper and olive, have regenerated.

One of the major challenges for the government is the increasing number of landless people who need alternative sources of income. There is also a strong push from the government to bring the farmers into the monetized economy through the development of markets. Local communities are being supported to build roads, and areas by these roads are being given to landless families to form small settlements, build houses and set up shops and markets. This is increasing the demand for wood to build houses. Often, arrangements are made for the new homeowners to buy and build their houses using eucalyptus. However, the demand for wood, particularly good quality wood which is durable and termite-proof, is very high. The preferred timber trees are juniper and olive. Usually the only places where juniper and olive trees of usable size are still found growing are in sacred forests. Despite the strong religious taboos, reports of old olive trees being cut to make new houses were brought to the attention of ISD and the local authorities in 2005. The threat to sacred forests was surveyed and it was found that there are many pressures being used by some local people to get access to the valuable timber in these forests. Although the BoARD, particularly the Department for Natural Resources and Environmental Protection, had developed guidelines for protecting forests, these were not being fully implemented because the local judiciary and administration at both tabia and woreda levels were not recognizing the bylaws of local communities. Now, in 2009, bylaws are recognized and their enforcement is being supported by all levels of government in the region.

Farmers' experiences in the project

At the start of the project, farmers were reluctant to plant trees around their farms, thinking that they would attract birds and shade their crops. But after they saw the impact on the health of their animals of having more forage available, particularly from *Sesbania*, the small multi-purpose tree legume, and appreciated that the trench bunds and check dams stabilized the land, they readily planted more trees. Many farmers have started to also plant fruit trees, both around their homesteads and in rehabilitated gullies. There is now a steady demand for fruit tree seedlings by the farmers and new nurseries have been developed to meet this demand.

Farmers have started diversifying their production once the quality of their land has improved; for example, through innovative intercropping, as already described. Another farmer reintroduced a special barley variety called "demhay", which is used to make a popular snack of roasted grain called "qolo". Overall, the diversity of most crops has increased in all the communities in the project.

In Ziban Sas, after the grazing area was restored in 1998, the farmers asked for other improvements, particularly a pond for watering their animals as the community is on a small plateau and the animals had to be taken several kilometres down into the valley below to get water. The first pond was only moderately successful because it was small and leaked. Starting in 2004, a much larger pond was constructed and provided with a cement lining. It was completed in 2005. Helping communities to construct communal ponds has now become a regular feature in the project's work plan.

Farmers, development agents and experts, local administrations and ISD staff have identified the following as the positive effects of the Tigray Project:

- Crop yields are as good as, and often better than, those obtained by using chemical fertilizer.
- Agro-biodiversity is maintained and improved. For example, in 1996 the farmers of Ziban Sas were growing only a durum wheat-barley mixture called 'hanfets' and a little teff, but now other crops such as maize and faba bean are also grown. Obviously this will make their agricultural system more complex and thus more likely to be resilient to climate change.
- Both biomass and biodiversity increase in the areas protected from free range grazing, with many plant and animal species that had disappeared from the local ecosystems

returning; for example, Aardvark, which digs up termites. The improved vegetation cover protects the soil from erosion and provides good bee forage, helping the farmers and their ecosystems become more resilient to climate change.

o Weeds are reduced in fields where compost has been applied – weed seeds, pathogens and insect pests are killed by the high temperature in the compost pits, but earthworms and other useful soil organisms establish well. Weeds that do well on poor soils, such as wild oats (*Avena vaviloviana*) and striga (*Striga hermonthica*), are much reduced in soil fertilized with compost.

o Increased moisture retention capacity of a soil – if rain stops early, crops gown on soil treated with compost resist wilting for about two weeks longer compared to fields treated with chemical fertilizer. This is crucial during times of drought, which remains a recurring problem in many parts of Ethiopia, especially along the east.

o Plants grown with compost are more resistant to pests and diseases than crops grown with chemical fertilizer.

o Residual effect of compost – the positive effects of compost can remain for up to four years. The farmers have realized that, in contrast to chemical fertilizers, they do not need to apply compost each year as after adequate amounts of compost have been applied in one year, they can obtain good yields from their crops for the next two to three years without applying compost afresh.

o Farmers have been able to get out of debt from buying chemical fertilizer – the economic returns from making and using compost are positive as farmers have been able to stop buying chemical fertilizer, but they get even higher yields.

o Foods made from grain grown in fields fertilized with compost are said to have a better flavour than foods made from crops grown in fields treated with chemical fertilizer.

Each project site has its distinctive features and problems, and the outcomes achieved are also distinctive. However, in general for all project sites, the environment has been rehabilitated, food and feed production greatly improved, tree and grass cover as well as biodiversity returned, and soil protected from erosion. By 2008, the BoARD claimed that soil erosion in the region had been reduced by over 60 percent.

SCALING UP THE TIGRAY PROJECT
Institutional ownership

Since it started in 1996, the Tigray Project has been accepted by the agricultural professionals in the BoARD and implemented in partnership with the ISD in collaboration with the respective woreda administrations and agricultural experts and development agents that work with the local farming communities. The Project Officer was one of its experienced extension officers assigned to the project by the BoARD. In 2002, the Project Officer was retired and became a full time staff member of ISD, retaining strong working relations with the Bureau. Hence, ISD has deliberately not set up a separate system to implement activities. Instead, ISD facilitates some inputs, gives advice, runs workshops and makes careful follow-up with all its local actors.

Since 2003, ISD and the BoARD have also worked closely with the EPA in its Land Rehabilitation Project.

Steps in the scaling up process in Tigray

The Gu'emse community was the focus of the annual farmers' field day in October 1998, and this encouraged the senior members of the BoARD to promote the "sustainable agriculture package" as part of the extension programme to improve food security for the region. This was the start of scaling up of the project's activities. Efforts were made to extend the project's activities to 90 communities and have 2 000 farmers making compost by the year 2000. This was only partly successful mainly because local experts and development agents lacked the required training in compost making.

As part of the scaling up process, ISD published in 2002 a booklet on compost making written in Tigrinya (the regional language of Tigray) by Arefayne Asmelash (1994 EC/2002 GC), followed by a poster in Tigrinya published in 2007. Although the guidebook and poster have been distributed widely, farmers, development agents and local experts need to have practical training and visiting of farming communities where compost has been used successfully in order to feel confident that they themselves can make and use compost effectively. Therefore, as of 2005, ISD changed its training approach to include at least 50 percent participation by farmers as trainers-of-trainees (TOT) in the making and using of compost alongside the agricultural professionals. The farmers were charged with the responsibility of training a minimum of ten of their neighbours in making and using compost with the agricultural professionals doing the follow-up monitoring of the model farmers. Trainer farmers who fulfil their obligation are awarded certificates, while those who succeed in training many of their neighbours also receive tools. Through this approach, two outstanding farmers, one a woman and the other a man, have managed to train over 30 of their neighbours in making and using compost.

The interest and support of the woreda administration is also important in scaling up. In Tahtai Maichew, the woreda administration approached ISD and EPA in 2004 to help it bring the whole of the woreda into the Tigray Project. The scaling up started in July 2004 with a four-day workshop involving over 200 farmers. Each of the local experts introduced an aspect of the project, followed by two farmers describing how the activities in question had improved their livelihoods. Leading farmers from neighbouring communities were trained and encouraged to train other farmers in their neighbourhood, so that there could be a rapid increase in the numbers implementing the activities of the project, particularly restricting free range grazing and making and using compost. By 2008, 26 communities from all the 20 tabias of the woreda were implementing the project and the neighbouring woredas around the towns of Axum and Adua had also joined the project. The aim is to develop a rehabilitated "green belt" about 80 km wide, west to east, with all the farming communities practicing ecological agriculture across what was until recently a degraded landscape.

The new training approach of focusing on farmers as trainers has also led to rapid scaling up in the making and use of compost throughout the region as reflected in Figures 10 and 11.

FIGURE 10

Total recorded crop production in Tigray, 2003–2006

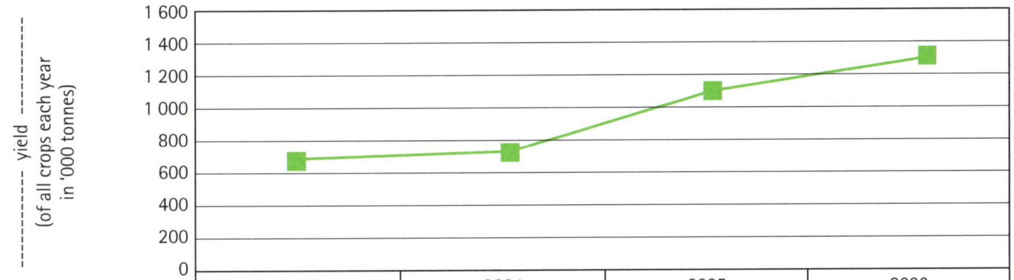

	2003	2004	2005	2006
crop yield in '000 tonnes	713.95	716.96	1 162.20	1 353.79

FIGURE 11

Total use of urea and DAP in Tigray, 1998–2005

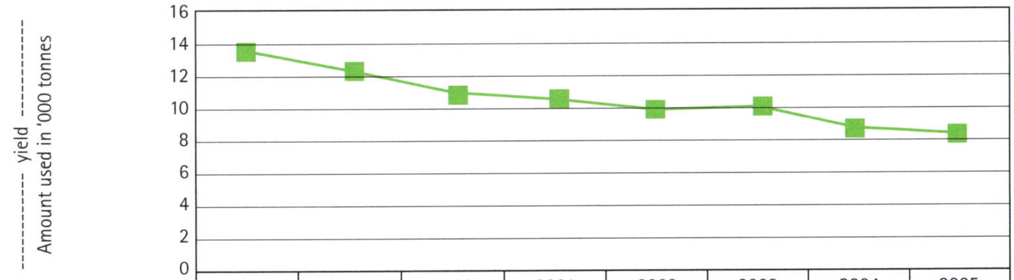

	1998	1999	2000	2001	2002	2003	2004	2005
crop yield in '000 tonnes	13.71	12.43	11.54	11.32	10.09	10.17	8.90	8.17

The official crop sample survey of Ethiopia for the crop year 2007–2008 showed that 86 percent of the nearly 700 thousand farmers in Tigray were using natural fertilizer on nearly 200 thousand hectares. Only 16 percent of the farmers had used chemical fertilizer on 48 thousand hectares. In the same crop year, the total production of cereals in Tigray reached 117 thousand tonnes with increases of between 12.7 and 15.9 percent for the major cereal crops. The implication from these data is that the bulk of this production was organic.

Steps in the scaling up process in other parts of Ethiopia
ISD's scaling up experience

The spread into Meqdela Woreda of South Wollo Zone in the drought prone eastern part of Amhara Region came about in 2004. Meqdela is the home area of one of the staff members of ISD, and he had talked about the Tigray Project when he had visited his family in Masha, the capital of Meqdela. In February 2004, six farmers, two local experts and two representatives from the local administration, including the woreda administrator, visited the Tigray communities, and four more participants came to a compost workshop in Axum in August 2004. When they returned to Masha, the local administrator encouraged the farmers to become local trainers with the aim of having 28 000 farmers in ten kebeles (parishes) in the woreda restricting free range animal grazing and making and using compost by the next growing season of 2005. Meqdela Woreda was visited in November 2005 and the impact of applying ecological agricultural practices was striking. As well as making compost, the farmers had started water harvesting, growing vegetables, and protecting hillsides from grazing. The local administration also changed a tree nursery into a nursery for producing fruit trees and forage plants. In 2005, Meqdela Woreda was awarded a prize by the Amhara Regional Government because of its efforts to rehabilitate its environment and improve the lifestyles of its farmers. It was the push by the local administrator for the farmers to become trainers in making and using compost that helped ISD adopt this approach in its own training programmes.

Scaling up has also been started in Gembichu of East Shewa Zone of the Oromiya Region, 45 km east of Addis Ababa. Two farmers had taken part in the big workshop in Axum in November 2001, and had been applying compost and fermented dung to their fields since. In September 2004, the farmers were invited to a workshop on seed saving where they challenged ISD to work with them. Compost making workshops have been held regularly since and one of the farmers formed a local neighbourhood group of 20 farmers committed to making compost. This group is also a Crop Conservation Association mandated to produce elite seed of farmers' varieties based on organic principles. In 2005, the administrator and experts of Gembichu Woreda requested ISD and EPA to include the woreda in their programme. Work has now been going on to get the whole woreda into an organic system of production.

EPA's scaling up experience

The first major event in the scaling up process into other regions of Ethiopia was a workshop held in Axum hosted jointly by ISD and the BoARD together with the Innovative Soil and Water Conservation Project of Mekelle University. This took place in November 2001 and representatives from all regions of Ethiopia as well as other NGOs working with farmers were invited. Over 40 farmers were also present. The papers presented included the first set of yield data from the use of compost in farmers' fields. The workshop also included field visits to the farms and farmers in Adi Nifas and to innovator farmers including one who had developed an ingenious lifting device to get water from a well 12 metres deep. He had used the water to develop a fruit and vegetable garden. This workshop stimulated experts from the EPA to develop a project for land rehabilitation with support from the United Nations Development Program (UNDP) under the Country Cooperation Framework Programme.

The Land Rehabilitation Project was launched in a workshop in Axum in October 2003. The aims of the project and the steps to achieve it were discussed and a field visit was carried out to Adi Nifas. The plan in the first phase of the project was to have communities from three woredas from each region, except Addis Ababa, start to implement the principles of the Tigray Project, but adapted to the local situation. The project was successful and a second five-year phase was launched, 2006 to 2010. This phase plans to include communities in 45 woredas in all regions and is based on each woreda developing and implementing a woreda environmental management plan based on their local community environmental management plans supported by appropriate local bylaws (EPA, 2010). The aim is to:

- Integrate environmental concerns in all development work at the local level;
- Ensure that agricultural production is based on ecological principles with reduced dependency on expensive external inputs;
- Fulfil Ethiopia's national obligations to the various global environmental conventions;
- Contribute towards the implementation of the Environmental Policy, and the Agricultural and Rural Development Policy of Ethiopia; and
- Improve rural livelihoods.

Between July 2006 and June 2008, communities in 27 woredas had initiated developing and implementing their respective environmental management plans. These woredas are in Afar, Amhara, Benshangul-Gumuz, Dire Dawa, Oromia, Southern Nations, Nationalities and Peoples, and Tigray Regions. By 2010, between 18 000 and 20 000 households will have benefited from this project, giving a total of around 100 000 beneficiaries.

The project works closely with the Extension Program of the Ministry of Agriculture and Rural Development as well as a number of non-governmental actors: local NGOs, the Small Grants Programme of the Global Environment Facility (GEF), the Nile Basin Initiative, the Global Mechanism of the United Nations Convention to Combat Desertification (UNCCD) and the Norwegian Development Fund.

In order to accommodate the challenge of providing training in the making and use of compost, ISD has developed a package of training materials and methodology for running compost making workshops as well as for follow-up.

The package includes the manual and an accompanying poster on making compost in Amharic, Oromiffa, Tigrinya, and English, a 30 minute video/DVD in Amharic, 30 annotated slides as a Power Point presentation, and a brochure describing how to organize and conduct a compost training workshop, and monitor the impact of its use in farmers' fields. EPA has supported ISD in getting the materials printed and distributed.

The experience from the Tigray Project has thus been expanded throughout Ethiopia by the Ministry of Agriculture and Rural Development through its extension system together with the EPA and ISD. Data gathered by the Ministry of Agriculture and Rural Development show that, in the year between the middle of 2007 and 2008, about 1.8 million hectares of farmland (about 16 percent of the total in the country), cultivated by about two million farming families (also about 16 percent of the total in the country), were fertilized with compost. Since a farm needs compost only every two to four years, this means that at least a third of the farmland in Ethiopia is organically fertilized. Towards the end of November 2008, a conference aimed at expanding the experience of the Tigray Project to other countries in Africa was convened by the African Union. The conference, which was held in Addis Ababa, was preceded by field visits to some of the Tigray Project sites near Axum in the Tigray Region.

Workshops and cross visits

The best person to convince a farmer to take up compost making is another farmer, and similarly, the best person to convince an expert or administrator of the value of the Tirgray Project is another person already involved in the project at the same responsibility level. For farmers, it is seeing that is believing, but this is also true for trained professionals. ISD has, therefore, invested time, energy and funds into developing good workshops and taking representatives from all levels in a woreda to meet their counterparts in another woreda where the project is being implemented. These are referred to as cross visits.

Through these workshops and cross visits, the development agents and local experts gain confidence to help the farmers make good compost and the relationship between them and the farmers changes from one of imposition to one of support and advice. All stakeholders also actively participate in evaluating the effectiveness of the activities of the project.

Compost workshops are now held regularly at the end of the growing season, i.e. from late August to early October, because this is the time when there is adequate green biomass available. Typically they last for three days and include a field visit as well as practical training and discussions. Therefore, the numbers of participants are kept below 50 and the practical work is done in groups of 10–15. Everyone is expected to physically contribute to filling up a compost pit. The discussions raise issues of what constitutes ecological agriculture and any special challenges faced by the farmers. Through these workshop opportunities, ISD has been able to hear and discuss the views of farmers on food security and on ecological agriculture, traditions in selecting and keeping seed for sowing, and preferred indigenous trees for animal feed and other uses. The workshops bring together farmers and local experts from different areas resulting in vigorous exchanges of views and experiences.

With the farmers being encouraged to become their own trainers, the technology can spread fast throughout a community. This is done by farmers committing themselves to train at least ten of their neighbours, and those trained also commit themselves to train other farmers. This can initiate a training chain so that all the farmers within one area make and use compost.

CONCLUSION

The experience in Tigray from the Tigray Project has been very positive, both in terms of helping farmers improve their livelihoods and in showing local experts, administrators and policy makers that environmental rehabilitation is basic to improving the productivity of crops and livestock. Use of ecologically sound organic principles can have very quick positive impacts on the productivity and well-being of smallholder farmers so that they do not necessarily have to face a conversion period of reduced yields as they change from chemical to organic production.

Most farmers, particularly those in degraded and drought prone areas, are not able to afford external inputs. For them, an organic production system offers a real and affordable means to break out of poverty and obtain food security.

It is important to bear in mind that although it may be external market interests that initially stimulate the development of a policy environment for organic agriculture, the benefits should be available to all members of society to build a healthy and food-secure future for Ethiopia.

While challenges undoubtedly still lie ahead, the positive experiences from Tigray and elsewhere in the country where farmers are using ecological/organic agriculture practices provide hope for Ethiopian farmers and improved food security for the country as a whole. Modified to suit local conditions for farmers in other parts of Africa, the lessons learnt from the Tigray Project could also bring substantial benefits to their local communities and the world of the future.

Considering the growing need for, and importance of, ecological agriculture, it is time that the research and development institutions of Ethiopia, Africa and the rest of the world take it seriously and breed varieties appropriate for it. Their approach should take advantage of the skills of smallholder farmers as participating partners in plant breeding, and of other methods to intensify ecologically sound crop management systems.

REFERENCES

Alemu Asfaw. 2001. The current depressed cereal prices: Reasons, impacts and its policy implications on future sustainable agricultural development (The case of maize belt areas, Western Ethiopia). Unpublished.

Alvares, F. 1961. *The Prester John of the Indies.* Hakluyt Society, London.

Arefayne Asmelash. 1994 EC/2002 GC. *Dekhuíi tefetro: intayn bkhemeyn* (Making compost: what it is, and how it is made). In Tigrinya. Tigray Bureau of Agriculture and Natural Resources and Institute for Sustainable Development, Addis Ababa.

Assefa Hailemariam. 2003. Population growth, environment and agriculture in Ethiopia. *In*: *The Symposium Proceedings, Population and development in Ethiopia: Now and in the future.* 7 June 2003. Walta Information Center, Addis Ababa.

Asseffa Abegaz. 2005. *Farm management in mixed crop-livestock systems in the Northern Highlands of Ethiopia.* Wageningen University and Research Center. (Ph.D. thesis)

Chaboussou, F. 1985. *Healthy crops: A new agricultural revolution.* Translated from French by M. Sydenham, G. Foley and H. Paul, and published in 2004 by Jon Carpenter Publishing, Charlbury and the Gaia Foundation, Brazil and London.

Clark, J. D. 1976. Prehistoric populations and pressures favouring plant domestication in Africa. *In* J.R. Harlan, J.M.J. de Wet and A.B.L. Stemler (eds.) *Origins of African plant domestication.* Mouton Publisher, The Hague.

CSA. 2002. *Ethiopia, Statistical Abstract 2002.* Central Statistics Authority. Addis Ababa.

Daniel Gemechu. 1977. *Aspects of climate and water budget in Ethiopia.* Addis Ababa University Press, Addis Ababa.

Edwards, S.B. 1991. Crops with wild relatives found in Ethiopia. *In* J.M.M. Engels, J.G. Hawkes and Melaku Worede (eds.) *Plant genetic resources of Ethiopia.* Cambridge University Press, Cambridge.

Edwards, S. 2003. *Natural fertilizer.* Institute for Sustainable Development, Addis Ababa.

Edwards, S., Arefayne Asmelash, Hailu Araya & Tewolde Berhan Gebre Egziabher. 2007. *Impact of compost use on crop yields in Tigray, Ethiopia. 2000–2006 inclusive.* FAO, Rome. Available at http://www.fao.org/documents/pub_dett.asp?lang=en&pub_id=237605 (accessed July 2009; verified 28 February 2010).

EMA. 1988. *National Atlas of Ethiopia.* Ethiopian Mapping Authority, Addis Ababa.

Engels, J.M.M. & J.G. Hawkes. 1991. The Ethiopian gene centre and its genetic diversity. *In* J.M.M. Engels, J.G. Hawkes and Melaku Worede (eds.) *Plant genetic resources of Ethiopia.* Cambridge University Press, Cambridge.

EPA. 2003. *The Federal Democratic Republic of Ethiopia – State of Environment Report for Ethiopia.* August 2003. Environmental Protection Authority, Addis Ababa.

EPA. 2010. A manual for the preparation of woreda and local community plans for environmental management for sustainable development. First produced in Amharic by the Environmental Protection Authority and translated into English with an introduction by Tewolde Berhan Gebre Egziabher. *In: Climate change and food systems resilience in sub-Saharan Africa.* FAO, Rome. (Accepted for publication).

FAO. 1986. *Highlands reclamation study of Ethiopia.* Final Report, Volume 1. FAO, Rome.

FDRE. 1989 EC/1997 GC. *Environmental Policy of the Federal Democratic Republic of Ethiopia.* Environmental Protection Authority with the Ministry of Economic Development and Cooperation, Addis Ababa.

FDRE. 2006. *Ethiopian Organic Agriculture System.* Negarit Gazeta Proclamation No.488/2006 approved by the Parliament on 8 March 2006.

FDRE. 2008. Federal Democratic Republic of Ethiopia: Population Census Commission 2008. *Summary and statistical report of the 2007 population and housing census: Population size by age and sex.* FDRE, Addis Ababa.

Hailu Araya. 2003. Seed security for food security: A study on Tigray Region, Ethiopia. Report commissioned by the African Biodiversity Network. Unpublished.

Hailu Araya & S. Edwards. 2006. *The Tigray experience: A success story in sustainable agriculture.* Environment and Development Series No. 4. Third World Network, Penang.

Hailu Araya & Hailemariam Gebrewahid. 2006. Report on the assessment of the situation of forests in the sacred places of Tigray Region. Unpublished.

Hunting Technical Services Ltd. 1976. *Tigrai rural development study. Annex 1 (Land and vegetation resources).* Hunting Technical Services Limited – Land and Water Resources Consultants in association with Sir M. Macdonald and Partners Consulting Engineers, London.

MoI. 2001. *The Government of the Federal Democratic Republic of Ethiopia – Rural Development Policy, Strategies and Guidelines.* Ministry of Information. Addis Ababa.

MoFED. 2002. *Ethiopia: Sustainable Development and Poverty Reduction Programme.* Federal Democratic Republic of Ethiopia – Ministry of Finance and Economic Development, Addis Ababa.

Nyssen J., Mitiku Haile, J. Naudts, N. Munro, J. Poesen, J. Moeyersons, A. Frankl, J. Deckers & R. Pankhurst. 2009. Desertification? Northern Ethiopia re-photographed after 140 years. *Sci. Total Environ.* 407(8): 2749-2755.

Nyssen J., J. Poesen, K. Descheemaeker, Nigussie Haregeweyn, Mitiku Haile, J. Moeyersons, A. Frankl, G. Govers, N. Munro & J. Deckers. 2008. Effects of region-wide soil and water conservation in semi-arid areas: The case of northern Ethiopia. *Z. Geomorph. N.F.* 52(3): 291–315.

Phillips, S. 1995. Poaceae (Gramineae). *In* I. Hedberg and S. Edwards (eds.), *Flora of Ethiopia and Eritrea*. Vol. 7. Addis Ababa and Uppsala.

Poncet, C.J. 1967. A narrative by Charles Jacques Poncet of his journey from Cairo into Abyssinia and back, 1698–1701. *In* W. Foster (ed.) *The Red Sea and adjacent countries at the close of the seventeenth century*. Kraus Reprint Ltd., Nendeln, Lichtenstein.

SSNC. 2008. *Ecological in Ethiopia: Farming with nature increases profitability and reduces vulnerability*. Swedish Society for Nature Conservation, Stockholm.

Tadesse Amare & Asferachew Abate. 2008. *An assessment of pesticide use, practices and hazards in the Ethiopian Rift Valley in 2007*. Institute for Sustainable Development, Ethiopia and Pesticide Action Network, UK.

Tewolde Berhan Gebre Egziabher, S. Edwards & Hailu Araya. 2004. *Ecological agriculture with small holder farmers in Ethiopia*. Institute for Sustainable Development, Addis Ababa.

UNDP. 2004. *Human Development Report 2004*. Available at http://hdr.undp.org/en/reports/global/hdr2004/ (accessed July 2009; verified 28 February 2010).

WRI, UNEP, UNDP & World Bank. 1998. *1998–99 World Resources: A guide to the global environment*. Oxford University Press, Oxford.

ADOPTION OF ORGANIC FARMING TECHNOLOGIES: EVIDENCE FROM A SEMI-ARID REGION IN ETHIOPIA

Menale Kassie,
Precious Zikhali,
Kebede Manjur and
Sue Edwards

CONTENTS

TABLES

FIGURES

ABOUT THE AUTHORS

Menale Kassie is an agricultural economist based in the Department of Economics, University of Gothenburg, Sweden, who works with the Environmental Economics Policy Forum for Ethiopia, based in Addis Ababa.

Precious Zikhali is with the Centre for World Food Studies, VU University, Amsterdam.

Kebede Manjur is in the Tigray Agricultural Research Institute (TARI), and works at Alamata Agricultural Research Center, Tigray, Ethiopia.

Sue Edwards is the Director of the ISD, Addis Ababa, Ethiopia, and has been co-editor of the eight-volume "Flora of Ethiopia and Eritrea" since 1984. She is a taxonomic botanist, teacher and science editor by profession.

EDITORIAL NOTES

This paper was first published in January 2009 as Environment for Development (EfD) Discussion Paper 09-01, a joint publication of the Environment for Development Initiative and Resources for the Future (www.rff.org).

INTRODUCTION

Sustainable agriculture, which can lead to formally recognized organic farming, can be broadly defined as an agricultural system involving a combination of sustainable production practices in conjunction with the discontinuation or the reduced use of production practices that are potentially harmful to the environment (D'Souza *et al.*, 1993). More specifically, the Food and Agriculture Organization of the United Nations (FAO) argues that sustainable agriculture consists of five major attributes: it conserves resources, it is environmentally non-degrading, technically appropriate, economically and socially acceptable (FAO, 2008). In practice, sustainable agriculture uses fewer external off-farm inputs such as purchased fertilizers and employs locally available natural resources, as well as purchased inputs, more efficiently (Lee, 2005).

Stubble tillage and compost are two examples of sustainable agriculture practices appropriate for organic farming.

Stubble tillage is a type of conservation farming in which farmers leave all or most of the stubble from the previous crop in place and turn it into the soil by ploughing soon after harvest to avoid grazing by livestock so that the organic material can decompose before the next cropping season. Conservation farming seeks to achieve sustainable agriculture through minimal soil disturbance with a permanent soil cover and crop rotation. The potential benefits from conservation farming lie not only in conserving but also in enhancing the natural resources, such as increasing soil organic matter, without sacrificing yield levels. This makes it possible for fields to act as a sink for carbon dioxide (i.e. for carbon sequestration), increase soil water-holding capacity, and reduce soil erosion. It also cuts direct production costs for the farmer and reduces the emission of greenhouse gases associated with mechanized farming (FAO, 2008). That conservation farming can address such a broad set of farming constraints makes it a widely adopted component of sustainable agriculture (Lee, 2005). However, conservation farming that is based on zero tillage is usually associated with the use of synthetic herbicides, which is not appropriate for or acceptable in organic farming.

The use of compost is part of an organic agriculture system that emphasizes maximum reliance on renewable farm and other local resources. Compost is an organic fertilizer that has the advantage that it improves soil structure and aeration, increases the soil's water-holding capacity and stimulates healthy root development (Twarog, 2006). Thus, both stubble tillage and compost may be appealing options for enhancing productivity with resource-poor farmers, especially in developing countries.

The agriculture sector in Ethiopia is the most important sector for sustaining growth and reducing poverty. It accounts for 50 percent of gross domestic product (GDP), 88 percent of export value and is a source of employment for more than 85 percent of the country's population of more than 70 million (Deressa, 2007). However, lack of adequate nutrient supply, the depletion of soil organic matter and soil erosion are major obstacles to improving agricultural production (Grepperud, 1996; Kassie *et al.*, 2008a). The key to a sustained increase in agricultural production is to improve the productive capacity of farmers, which can be achieved for example through improved farming management practices and more effective use of existing and new technologies. Inorganic fertilizer remains the main yield-augmenting technology being aggressively promoted by the government and other institutions. Despite this promotion, inorganic fertilizer adoption rates remain very low. Until recently, only 37 percent of farmers used inorganic fertilizer, and application rates remained at or below 16 kg/hectare of nutrients (Byerlee *et al.*, 2007), which is far below the recommended rate of 110 kg/hectare of nutrients.[1] In addition to low application rates, there is evidence suggesting a pull back from using fertilizer (EEA/EEPRI, 2006). Escalating prices and production and consumption risks (that is, variability of agricultural production due to rainfall variability which causes consumption variability in the absence of consumption smoothing mechanisms) have been cited as among the factors limiting the use of inorganic fertilizers in Ethiopia (Dercon and Christiaensen, 2007; Kassie, *et al.*, 2008b).

---

[1]   The recommended rates are 100 kg per hectare of diammonium phosphate (DAP) and 50 kg per hectare of urea.

Thus, given the aforementioned challenges to inorganic fertilizer adoption, a key policy intervention for sustainable agriculture is to encourage adoption of farming technologies that rely, to a greater extent, on renewable farm and other local resources. Organic farming practices, such as the use of compost and stubble tillage, are among such technologies. The water retention characteristics of these technologies (Twarog, 2006) make them especially appealing in water-deficient farming areas, such as the Tigray Region in northern Ethiopia. In addition to reducing natural risks, organic farming practices enable poor farmers to minimize the financial risk of buying chemical fertilizer on credit and - given that compost and stubble tillage are available when needed - alleviate the prevailing problem of late delivery of chemical fertilizer (Hailu and Edwards, 2006).

Since 1998, Ethiopia has included stubble tillage and use of compost as part of its extension package to reverse extensive land degradation (Edwards *et al.*, 2007; Sasakawa Africa Association, 2008). There exists ample evidence to show that use of compost and stubble tillage can result in higher and/or comparable yields compared to chemical fertilizer (Edwards *et al.*, 2007; Hailu and Edwards, 2006; Hemmat and Taki, 2001; Kassie *et al.*, 2008b; Mesfine *et al.*, 2005; Sasakawa-Global 2000, 2004; UNCTAD and UNEP, 2008). This implies that these two organic farming technologies can create a win-win situation, where farmers are able to reduce direct production costs, improve environmental benefits, and, at the same time, increase their crop yields.

Although numerous studies have been carried out in Ethiopia on the adoption and subsequent economic impacts of chemical fertilizer, improved seed and physical soil and water conservation structures, some of which have been published (e.g. Croppenstedt *et al.*, 2003; Dercon and Christiaensen, 2007; Hagos, 2003; Kassie *et al.*, 2008a; Shiferaw and Holden, 1998), no study, to the best of our knowledge, has investigated the determinants of the adoption of compost use and conservation farming in the form of stubble tillage by farmers in Ethiopia (Neill and Lee, 2001). The objective of this study was to look at how socio-economic and biophysical characteristics determine adoption of compost and chemical fertilizer use, and stubble tillage in the semi-arid region of southern Tigray, Ethiopia. By identifying significant characteristics associated with the adoption of these sustainable agriculture practices, information to support formulation of policies and strategies that promote the adoption of these practices can be put forward. In addition, a dataset containing information on the use of compost and chemical fertilizer in relation to grain yields was used to perform a stochastic dominance analysis with the aim of examining whether adoption of these technologies had any impacts on productivity. By showing the importance of organic practices in enhancing productivity, we can validate the need to further investigate the factors that condition their adoption by farmers.

The results reveal a clear superiority with the use of compost, compared to chemical fertilizers, when it comes to crops yields. Regarding the determinants of adoption decisions, it was found that, while there is heterogeneity in the factors affecting the choice to use compost and/or stubble tillage, both plot and household characteristics influenced adoption decisions. Interestingly, we also found evidence that the impact of gender on technology adoption is technology-specific, while the significance of plot characteristics indicated that the decision to adopt a given technology is location-specific.

THE ANALYTICAL AND ECONOMETRIC FRAMEWORK

The analysis began by using a non-parametric technique, the stochastic dominance analysis, to assess how the use of compost and chemical fertilizer impacts crop production. Stochastic dominance analysis is used to compare and rank distributions of alternative risky outcomes according to their level and dispersion (riskiness) of returns (Mas-Colell *et al.*, 1995). The comparison and ranking is based on cumulative density functions using the entire density of yield data, rather than averages, in the yield data set.

It was posited that both plot and household socio-economic characteristics influence the decision to adopt technologies. Plot characteristics condition the decision to adopt one specific technology over another by their impact on the increment of plot profit or the productivity impact derived from participation. Farmers' socio-economic characteristics and preferences, on the other hand, might result in different adoption decisions, even when plots have similar characteristics. Accordingly, the maximizing of a technology's utility for a farmer forms the basis of our econometric model and estimation strategy. The framework assumes that if adoption of several farming practices is possible, it is expected that, in deciding to adopt one or more practices, a farmer compares the indirect utility values associated with each practice or a combination of practices.

Consequently, to study the ith farmer's choice, random utility models were postulated, each being associated with the jth choice of farming practice, such that:

$$V_{ij} = \mathbf{X_i}'\mathbf{\beta_j} + \varepsilon_{ij} \; , \tag{1}$$

where V_{ij} is the indirect utility level which the ith farmer associates with the jth farming practice, and $\mathbf{X_i}$ is a vector describing the farmer's socio-economic characteristics, as well as plot characteristics. The vector of parameters to be estimated is denoted by $\mathbf{\beta}$, while ε is the stochastic error term. Given the two organic farming practices, use of compost and stubble tillage, this provided four feasible choices available to the farmer. These were classified such that $j = 0$ if neither of the two practices is adopted, $j = 1$ if compost is adopted, $j = 2$ if stubble tillage is adopted, and $j = 3$ if both compost and stubble tillage are adopted.

Given a dummy variable, d_{ij}, to capture the choice of the ith farmer regarding the jth farming practice, the farmer's decision rule then becomes:

$$\begin{cases} d_{ij} = 1 \\ \\ d_{im} = 0 \, \forall m \neq j \end{cases} \Leftrightarrow (V_{ij} > V_{im} \quad \forall m \neq j) \; . \tag{2}$$

To make the econometric model operational, it was assumed that the disturbances of the different combinations are independent and identically distributed with the Gumbel cumulative distribution function, which implies that the probability of choosing the *j*th combination becomes (Greene, 1997):

$$P_{ij} = \Pr(d_{ij} = 1) = \frac{\exp(\mathbf{X}_i'\boldsymbol{\beta}_j)}{\displaystyle\sum_{m=0}^{j} \exp(\mathbf{X}_i'\boldsymbol{\beta}_m)} \qquad (3)$$

This is the multinomial logit model, characterized by the independence of irrelevant alternatives, which implies that from equation (3) the following expression can be derived:

$$\frac{P_{ij}}{P_{im}} = \exp(\mathbf{X}_i'(\boldsymbol{\beta}_j - \boldsymbol{\beta}_m)) \qquad \forall\, m \neq j \qquad (4)$$

a condition which holds for whatever subsets of eligible combinations include *j* and *m*. Given that the model is based on the difference of expected utility levels in each pair of combinations, it was assumed that $\boldsymbol{\beta}_0 = 0$ to solve the problem of the indeterminacy, which could complicate the estimation of the model (Greene, 1997). The maximum likelihood procedure was used to solve the model.

THE DATA AND DESCRIPTIVE STATISTICS

This study benefited from two datasets. The first was a cross-sectional dataset collected in 2006 in the Ofla Woreda (district) of South Tigray Region that was used to analyse the determinants of adoption of compost, stubble tillage and chemical fertilizer. It included a random sample of 130 households across five villages from which observations from 355 plots were collected. Due to missing values for some of the explanatory variables the number of observations used in the final sample is for 348 plots. In addition to information on adoption of compost, stubble tillage and chemical fertilizer, the dataset contained household and plot characteristics, plus indicators of access to infrastructure which, based on economic theory and previous empirical research, were included in the analysis. The descriptive statistics of the variables used in the regression analysis are presented in Table 1.

TABLE 1

Descriptive statistics of variables from Ofla Woreda used in the analysis

VARIABLES	DESCRIPTION	MEAN	STD. DEVIATION
Dependent variables			
Stubble tillage	Plots received stubble tillage (1 = yes; 0 = otherwise)	0.368	0.483
Compost	Plots received compost (1 = yes; 0 = otherwise)	0.170	0.376
Chemical fertilizer	Plots received chemical fertilizer (1 = yes; 0 = otherwise)	0.236	0.425
Wheat grain yield with compost	Wheat grain yield in tonne per hectare	7.480	4.020
Wheat grain yield with chemical fertilizer	Wheat grain yield in tonne per hectare	5.080	2.470
Wheat grain yield without compost and chemical fertilizer (control yield)	Wheat grain yield in tonne per hectare	3.680	1.870
Barley grain yield with compost	Barley grain yield in tonne per hectare	7.047	2.725
Barley grain yield with chemical fertilizer	Barley grain yield in tonne per hectare	5.583	2.380
Barley grain yield without compost and chemical fertilizer (control yield)	Barley grain yield in tonne per hectare	3.217	1.960
Teff* grain yield with compost	Teff grain yield in tonne per hectare	6.790	5.000
Teff grain yield with chemical fertilizer	Teff grain yield in tonne per hectare	5.228	2.665
Teff grain yield without compost and chemical fertilizer (control yield)	Teff grain yield in tonne per hectare	3.450	2.780

VARIABLES	DESCRIPTION	MEAN	STD. DEVIATION
Explanatory variables			
Socio-economic characteristics			
Male	Sex of household head (1 = male; 0 = female)	0.822	0.379
Age	Age of household head	41.084	12.902
Dependents	Number of economically inactive household members	2.504	1.642
Household labour	Number of economically active household members	2.275	0.869
Illiterate (cf.)	Household head has no education (1 = yes; 0 = otherwise)	0.529	0.489
Religious education	Household head has religious education (1 = yes; 0 = otherwise)	0.114	0.319
Formal education	Household head has formal education (1 = yes; 0 = otherwise)	0.357	0.468
Farmers' organizations	Membership in farmers' organization (1 = yes; 0 = otherwise)	0.134	0.334
Extension	Household extension contact (1 = yes; 0 = otherwise)	0.636	0.469
Livestock	Household livestock holding, in Tropical Livestock Units (one TLU is equivalent to an animal that weighs 250 kg)	2.774	2.522
Farm size	Total farm size, in hectares	0.777	0.666
Market distance	Distance from residence to the district market, in hours	2.054	1.971
Plot characteristics			
Ownership**	Whether the householder owns the plot (1 = yes; 0 = otherwise)	0.718	0.450
Distance	Distance from residence to the plot, in minutes	0.626	0.629
Flat to moderate slope	Farmer's perception: plot is of flat to moderate slope (1 = yes; 0 = steep slope)	0.305	0.461
Fertile soil	Farmer's perception: plot is of fertile soil (1 = yes; 0 = infertile)	0.342	0.475
Black soil	Predominantly black soil (1 = yes; 0 = otherwise)	0.388	0.488
Deep soil (cf.)	Farmer's perception: plots with deep soil depth (1 = yes; 0 = otherwise)	0.379	0.486
Moderately deep soil	Farmer's perception: plots with moderately deep soils (1 = yes; 0 = otherwise)	0.241	0.429
Shallow soils	Farmer's perception: plots with shallow soil depth (1 = yes; 0 = otherwise)	0.379	0.486
Degradation	Plot perceived as being degraded (1=yes; 0=otherwise)	0.359	0.480
Number of plot observations			348

*   Teff (*Eragrostis tef*) is a cereal crop with very small grains that is the main ingredient of a fermented pancake-like bread called injera.

**   There is no private land ownership in Ethiopia, hence "ownership" indicates that the household head has a written legal leasehold to the land.

Source: Author's own calculations

Around 17, 36.8, and 23.6 percent of the plots used compost, stubble tillage, and chemical fertilizer, respectively. Regarding the householders' perceptions of the impacts of using compost and stubble tillage, about 40 and 74 percent of the adopters, respectively, perceived positive impacts of these technologies on soil fertility; about 20 and 42 percent, respectively, believed that these technologies reduced soil erosion; and 32 and 69 percent, respectively, believed that these technologies were labour intensive. The data also revealed that those farmers who adopted use of compost had more livestock compared to those who adopted stubble tillage. Conversely, farmers that adopted stubble tillage had larger farms (average 1.30 ha) compared to those who adopted using compost (average 1.03 ha). It was assumed that those farmers with larger farms who adopted stubble tillage could expect to produce more straw for livestock feed and could afford to plough in their stubble to increase soil fertility.

The fact that the first dataset did not include production data precluded our use of this dataset to analyse how adoption of these technologies impacted crop production. To achieve this objective, we employed a second dataset to conduct a stochastic dominance analysis. It was a cross-section time series of farm-level production data collected between 2000 and 2006 by the Institute for Sustainable Development (ISD)[2]. The dataset does not have any information on plot and household characteristics. ISD's primary objective for collecting these data was to investigate the impact of compost on crop production and soil fertility. The dataset covered eight districts and 19 villages in the Tigray Region, including the Ofla district from where the data used in the econometric analysis was collected. Of the 19 villages, 17 are located in drought-prone areas of the southern, eastern and central zones of Tigray. ISD collected agronomic data (grain and straw yields) for 11 crops from 974 plots. The FAO crop-sampling method was used to collect yield data from the plots which had received compost, chemical fertilizer and no inputs (control plots).

---

2   ISD promotes ecological agriculture in Ethiopia. It also has the responsibility of providing information and training on making compost and its application, and recording grain and straw yield data during harvest in collaboration with the local agricultural extension workers and farmers.

Three one-metre-square plots were harvested from each field to reflect the range of conditions of the crop. All of the crop management practices, including the amount of compost and fertilizer application, were decided by the farmers themselves as ISD did not set the level of inputs to be used by the farmers. The average-per-hectare cost of applying compost in 2007 was about ETB[3] 307 whereas commercial fertilizer (DAP and urea) is about ETB 497 (Hailu, 2010). The major cost component of compost is a labour cost, which accounts for 55% of the total cost (Michaele, 2005. Unpublished).

ESTIMATION RESULTS

In this section, the results obtained from stochastic dominance analysis and the multinomial logit adoption model are discussed.

Stochastic dominance analysis

For the purposes of this analysis, the three crops (wheat, barley, and teff) most widely grown by the farmers were used in order to compare yield distributions obtained from plots using compost, chemical fertilizer, and no inputs (control). The outcome variable is the physical grain yield (tonne/hectare) of the respective crops. Figures 1–3 show cumulative density functions for yields obtained from compost, chemical fertilizer and control plots.

As illustrated in the figures, for all crops the yield cumulative distribution with compost is entirely to the right of the chemical fertilizer and control yield distributions, indicating that yield with compost unambiguously holds first-order stochastic dominance over chemical fertilizer and control plots. That is, compared to control plots and plots that used chemical fertilizer, plots with compost gave significantly higher yield levels. Yield distribution of plots with chemical fertilizer

---

3   ETB = Ethiopian birr. 1 USD = 9.50 ETB as at December 2008.

FIGURES 1-3

Stochastic dominance analysis of the impact of compost on crop productivity in Tigray Region, Ethiopia, 2000-2006 inclusive

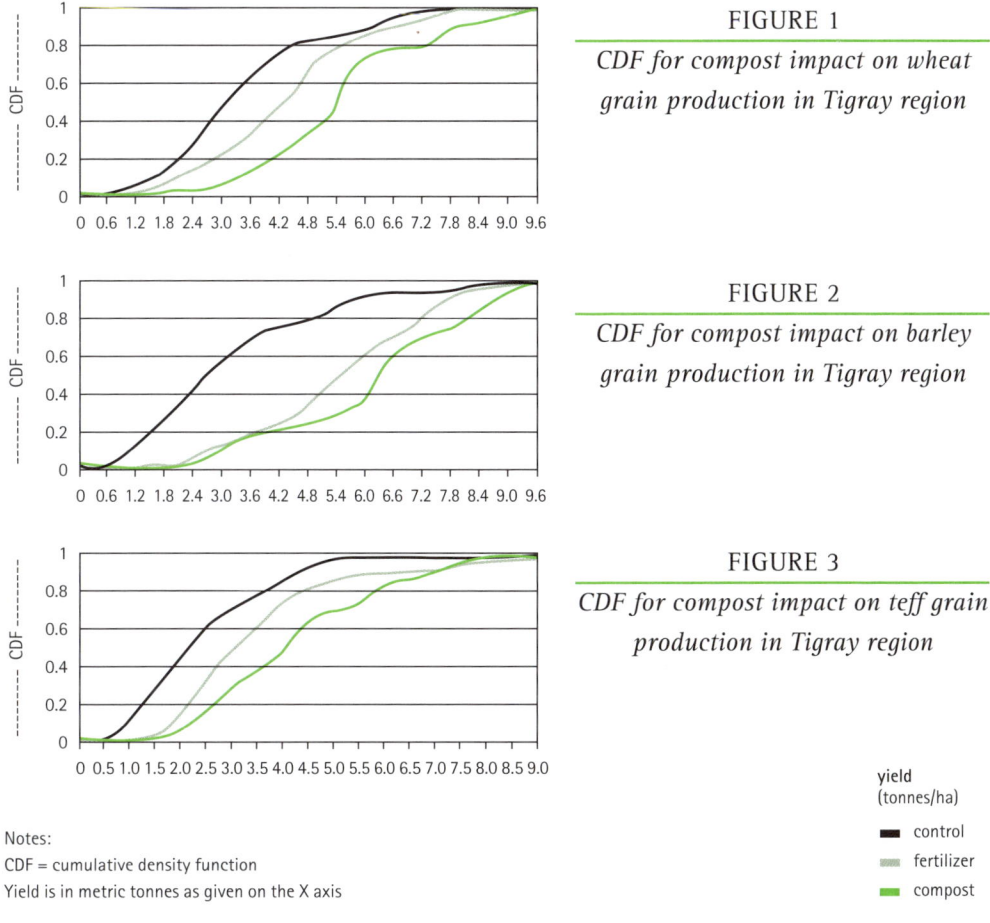

FIGURE 1

CDF for compost impact on wheat grain production in Tigray region

FIGURE 2

CDF for compost impact on barley grain production in Tigray region

FIGURE 3

CDF for compost impact on teff grain production in Tigray region

yield
(tonnes/ha)

■ control
▨ fertilizer
■ compost

Notes:
CDF = cumulative density function
Yield is in metric tonnes as given on the X axis

dominated yield distributions of control plots, i.e. plots with no input. However, crop production can also be influenced by plot and household characteristics apart from adoption of farming practices, for example, the gender of the household head (see later discussion).

The non-parametric Kolmogorov-Smirnov statistics test for first-order stochastic dominance (or the test for the vertical distance between the two cumulative density functions) also confirmed this result (see Table 2 below).

TABLE 2

Kolmogorov-Smirnov statistics test for first-order stochastic dominance

CROP	TREATMENTS		
	Compost versus control	Compost versus chemical fertilizer	Chemical fertilizer versus control
Barley	0.355 (0.000)***	0.192 (0.008)***	0.241 (0.000)***
Wheat	0.484 (0.000)***	0.384 (0.000)***	0.270 (0.000)***
Teff	0.591 (0.000)***	0.195 (0.003)***	0.396 (0.000)***

Note: *** significant at 1%

The foregoing analyses revealed an interesting finding: adoption of sustainable farming practices, such as the use of compost, is not inferior, in terms of its impact on yields, to the use of chemical fertilizers. In fact, as the results showed, use of compost can lead to significantly higher yields. While the use of profitability could be argued to be better than the use of yields (it could be the case that production plans with the best yields are not necessarily the most profitable), assuming the cost functions are correct, these results indicate that the use of compost could be more profitable than the use of chemical fertilizer. Other benefits from the use of compost could have been identified if these had been associated with measurements of environmental services and their long-term impacts on plot productivity. For example, unlike fields that receive chemical fertilizer in the current year, fields that received compost may not need a fertility enhancing input in the following year. Given these benefits, what constrains farmers from adopting such technologies? And, if they decide to adopt, what determines their choice of technology? The multinomial logit model was used to answer these questions. The results are discussed in the following section.

Multinomial logit model results

Table 3 below gives the multinomial logit-estimation results for the impact of both plot and socio-economic characteristics of the household on the decision to adopt a given farming practice. The base outcome is adopting neither of the practices, i.e. $j = 0$. This implies that the following discussion of the results focuses on the impact of the explanatory variables on a specific choice relative to no adoption. The model was tested for the validity of the independence of the irrelevant alternatives (IIA) assumptions, using the Hausman test for IIA. The test failed to reject the null hypothesis of the independence of the adoption of organic farming technologies, suggesting that the multinomial logit specification is appropriate for a model for adoption of organic technologies.

TABLE 3

Multinomial logit estimates

Variable	COMPOST		STUBBLE TILLAGE		BOTH	
	Coeff.	Std. error	Coeff.	Std. error	Coeff.	Std. error
Socio-economic characteristics						
Male household head	-1.99**	0.79	19.60***	0.92	2.21**	1.11
Age	-0.01	0.03	-0.06***	0.02	-0.06***	0.02
Dependents	-0.23	0.18	-0.20**	0.10	-0.17	0.12
Household labour	-0.14	0.42	0.40**	0.20	0.36	0.24
Religious education	0.39	1.06	-0.53	0.67	-0.31	0.76
Formal education	1.41*	0.73	0.14	0.39	0.25	0.47
Farmers' organizations	0.90	0.77	1.46***	0.47	1.24**	0.58
Extension	1.95**	0.85	1.00**	0.40	1.09**	0.51
Livestock	0.20**	0.10	0.05	0.05	0.06	0.07
Farm size	0.31	0.42	0.54***	0.20	0.39	0.25
Market distance	0.16	0.16	0.06	0.09	-0.07	0.12
Plot characteristics						
Ownership	1.38*	0.78	0.41	0.34	1.29**	0.50
Distance	0.64	0.44	0.51*	0.26	0.43	0.33
Flat or moderate slope	-1.26	0.78	-0.74**	0.37	-1.35***	0.52
Fertile soil	0.56	0.62	0.20	0.36	-0.16	0.45
Black soil	-0.57	0.61	0.65*	0.36	0.25	0.42
Moderately deep soil	-0.68	0.86	0.38	0.41	-0.74	0.57
Shallow soil	0.52	0.71	0.50	0.43	0.40	0.50
Degradation	-0.32	0.64	0.14	0.34	0.02	0.42
Number of observations	348					
Pseudo R2	0.23					
LR chi2 (54)	168.60***					
Log likelihood	-287.18					

Notes: Base outcome = no adoption; * significant at 10%; ** significant at 5%; *** significant at 1%

The results suggest that both socio-economic and plot characteristics are significant in conditioning the householders' decisions to adopt sustainable agricultural production practices. While there is heterogeneity with regard to factors influencing the choice to adopt compost and/or conservation tillage, these results suggest that significant determinants of adoption can be broadly classified into social characteristics of household head, labour availability, access to information, wealth and plot characteristics, which include whether or not the householder "owns" the plot.

There is a heterogeneous impact of the gender of the head of the household on adoption decisions regarding the two practices. Specifically, households with a male head are less likely to adopt the use of compost, while they are more likely to either adopt stubble tillage or combine it with the use of compost. While some researchers have found that male-headed households are more likely to adopt sustainable agricultural technologies (Adesina *et al.*, 2000), our results underscored the need to avoid generalizing the impact of gender on unspecified farm technology adoption, emphasizing that the impact of gender on technology adoption is technology-specific. In this area, it seems male-headed households had a comparative advantage in using stubble tillage, while female-headed households enjoyed an advantage in the use of compost. This is probably linked to the cultural constraint that prevents women from handling plough oxen and constraints on labour availability at harvest time when stubble tillage should be done. Still with the characteristics of the household head, there was a negative and significant impact of age on the likelihood of adopting stubble tillage, as well as combining it with compost. This could suggest that younger farmers are more likely to try innovations and, in addition, they might also have a lower risk aversion and longer planning horizon to justify investments in technologies whose benefits are realized over time. The results also suggest the need to develop gender- and age-specific technologies instead of blanket recommendations of technologies, regardless of the characteristics of the farmers, to encourage adoption of sustainable agricultural practices.

Labour issues seem to be of more concern in the decision to adopt stubble tillage. Specifically, the probability of adopting stubble tillage, relative to no adoption, increased with the number of household members who actively provided farm labour. This is in line with the results of the descriptive statistics, where about 69 percent of those farmers that adopted stubble tillage reported that it was labour intensive. This is not surprising because stubble is tilled during crop harvesting, which is one of the peak periods for agricultural labour. This underscores the importance of labour availability to technology adoption, consistent with findings by Caviglia and Kahn (2001) and Shiferaw and Holden (1998). In such circumstances, it is important to consider strengthening the existing local labour-sharing mechanisms. On the other hand, this probability declined with the number of dependents in the household, capturing the intuitive expectation that the time spent caring for dependents shifts labour away from adoption activities.

Access to information on new technologies is crucial to creating awareness and attitudes towards technology adoption (Place and Dewees, 1999). In line with this, access to agricultural extension services, indicated by whether or not the household head had contact with an extension agent, impacted the adoption of all technology choices positively. Contact with extension services gives farmers access to information on innovations, advice on inputs and their use, and management of technologies. In most cases, extension workers establish demonstration plots where farmers get hands-on learning and can experiment with new farm technologies.

Another indicator of information that shapes management skills is the amount of formal education (as opposed to no education at all), which increases the probability of using compost relative to not adopting any sustainable agriculture practice at all. This suggests that using compost is relatively knowledge-intensive and, thus, that management skills are crucial in its adoption. It has been argued that farmers' associations and unions constitute important sources of information for farmers (Caviglia and Kahn, 2001). These results confirm this: a householder's membership in at least one farmers' organization significantly increased the likelihood that the farmer would practice stubble tillage, as well as the likelihood that the farmer would use both compost and stubble tillage. These results underscore the role of public policy supporting farmers' organizations in encouraging the adoption of sustainable agricultural practices.

In addition, livestock ownership, i.e. fewer animals, limited the adoption of compost, while a householder's size of landholdings limited the adoption of stubble tillage (as well as combining the two practices). This suggests that poverty significantly limits technology adoption. Wealth intuitively affects adoption decisions since wealthier farmers have greater access to resources and may be better able to take risks. It must be acknowledged, however, that the wealth measures used might be confounded with other factors related to adoption. For example, using livestock ownership as an indicator of wealth may be compromised by the fact that oxen provide draft power as well as manure (which, as organic matter, is a component of compost). Furthermore, as the data show, farmers who use compost have more livestock compared to those who adopt stubble tillage. This result could imply that the opportunity cost of crop residues is smaller for stubble tillage adopters than compost adopters. The size of total landholdings, on the other hand, although measuring farmers' wealth, could also suggest economies of scale in production using stubble tillage, as well as indicating the social status of the household head – both of which could influence the farmers' ability to obtain credit. All the same, these results suggest that policies to alleviate poverty and increase crop productivity among farmers will impact the adoption of sustainable agricultural practices positively.

The results revealed that plot ownership has a positive impact on adoption decisions consistent with Arellanes (1994) for adoption of minimum tillage in Honduras and Neill and Lee (2001) for adoption of cover crops in Honduras. This is in line with the theory of Marshall's disincentive hypothesis where input use on rented land, particularly on sharecropped land, is found to be lower than on owned land because of the disincentive effect of output sharing (Kassie and Holden, 2007). Given the fact that the benefits from investing in both compost and stubble tillage accrue over time, this inter-temporal aspect implies that secure land access or tenure will impact adoption decisions positively. Distance of plots from residence positively and significantly influences adoption of stubble tillage. This could be reflecting the fact that, for example, crop residues in plots that are farther away from the homestead are less accessible to be used as livestock feed. This would then mean that farmers might prefer to use crop residues in these plots for stubble tillage instead. Alternatively, plots close to the homestead could be relatively fertile compared to distant plots because homestead plots may benefit from household refuse and other soil fertility enhancing materials (e.g. ash).

Sustainable agricultural systems are intuitively site-specific (Lee, 2005). This is further confirmed by the finding that plot characteristics influence the decision to adopt stubble tillage, as well as to combine it with the use of compost. In particular, the likelihood of householders choosing to practice stubble tillage declined with the perceived increase in the slope of the plots. This could reflect the fact that plots with steeper slopes are more prone to soil erosion, which necessitates adoption of farming techniques, such as stubble tillage, since they are meant to mitigate soil erosion and subsequent nutrient losses. The plot slope impacted the decision to combine the use of compost and conservation tillage in a similar way. It also found that conservation tillage is more likely to be practiced on plots with predominantly black soils, indicating the role of soil type and quality in influencing adoption decisions. Interestingly, plot-specific characteristics did not seem to impact the decision to adopt only the use of compost. These results imply that, for sustainable agricultural practices to be successful, they must address site-specific characteristics since these condition the need for adoption as well as the type of technology adopted.

CONCLUSION

In this paper, plot-level data from the semi-arid region of Tigray, Ethiopia, have been used to investigate the factors influencing farmers' decisions to adopt sustainable agriculture practices, with a particular focus on the adoption of stubble tillage and compost. In addition, stochastic dominance analysis is used to examine and compare the productivity gains of sustainable farming practices and chemical fertilizers.

While there is heterogeneity with regard to factors that influence the choice of any of the two practices, these results underscore the importance of both plot and household characteristics in adoption decisions. The findings imply that public policy can affect the adoption of sustainable agriculture. Specifically, the significant and positive impact of access to information, and the quality of that information, indicates that public policies aimed at improving access to information will help promote adoption of organic farming practices.

Moreover, it is shown that such public policies should take into consideration the influence of gender differences, particularly as these are linked to culture and differential labour demands between sexes, in adoption of different technologies. There is also evidence that the age of the household head (whether affecting aversion to risk and/or life-cycle dynamics) will have a differential impact on adoption, depending on the type of technologies. Availability of household labour also conditions the choice of technology adopted, given that the labour requirements differ from technology to technology as well as the timing of the application of the technology. Thus, public policy should factor in the impact of these socio-economic characteristics.

Security of access to land has a significant influence on a farmer's decision whether or not to adopt a technology, and its impact varies from technology to technology. Policies should strive to ensure that the drive to give farmers secure tenure rights are not undermined by other decisions on land use that could undermine the security of these tenure rights.

The viability of the agricultural production systems in Ethiopia, as in many areas in developing countries, is highly constrained by degraded soils and increasing lack of reliability in rainfall resulting from climate change. These problems are further compounded by an increasing population that is not accompanied by improvements in local technologies and land management practices. Even though efforts by the government are resulting in a larger proportion of the farmers making use of chemical fertilizers, the sustainability of this approach to increasing crop productivity and food security is at risk because of escalating prices for fertilizer and transport to bring the crops to market. Given these constraints, it can be argued that sustainable agricultural practices that emphasize the management capacity and self-reliance of farmers to improve their productivity can create a win-win situation for both the local communities and the country as a whole. Farmers are able to reduce production costs by relying on renewable local resources that, at the same time, improve environmental services and increase yields.

REFERENCES

Adesina, A.A., D. Mbila, G.B. Nkamleu & D. Endamana. 2000. Econometric analysis of the determinants of adoption of alley farming by farmers in the forest zone of Southwest Cameroon. *Agriculture, Ecosystems and Environment* 80: 255–65.

Byerlee, D., D.J. Spielman, D. Alemu & M. Gautam. 2007. *Policies to promote cereal intensification in Ethiopia: A review of evidence and experience.* International Food Policy Research Institute Discussion Paper 00707. IFPRI, Washington DC.

Caviglia, J.L. & J.R. Kahn. 2001. Diffusion of sustainable agriculture in the Brazilian rain forest: A discrete choice analysis. *Economic Development and Cultural Change* 49: 311–33.

Croppenstedt, A., M. Demeke & M.M. Meschi. 2003. Technology adoption in the presence of constraints: The case of fertilizer demand in Ethiopia. *Review of Development Economics* 7(1): 58–70.

Dercon, S. & L. Christiaensen. 2007. *Consumption risk, technology adoption, and poverty traps: Evidence from Ethiopia.* World Bank Policy Research Working Paper 4527. World Bank, Washington DC.

Deressa, T.T. 2007. *Measuring the economic impact of climate change on Ethiopian agriculture: Ricardian appproach.* World Bank Policy Research Working Paper 4342. World Bank, Washington DC.

D'Souza G., D. Cyphers & T. Phipps. 1993. Factors affecting the adoption of sustainable agricultural practices. *Agricultural Resource Economics Review* 22(2): 159–65.

EEA/EEPRI. 2006. *Evaluation of the Ethiopian agricultural extension with particular emphasis on the participatory demonstration and training extension system (PADETES).* Ethiopian Economics Association/Ethiopian Economic Policy Research Institute, Addis Ababa.

Edwards, S., Arefayne Asmelash, Hailu Araya & Tewolde Berhan Gebre Egziabher. 2007. *Impact of compost use on crop production in Tigray, Ethiopia.* FAO, Rome Avaliable at http://www.fao.org/documents/pub_dett.asp?lang=en&pub_id=237605 (verified 24 March 2010).

FAO. 2008. Conservation agriculture. Available at http://www.fao.org/ag/ca/ (accessed 16 December 2008; verified 24 March 2010).

Greene, W.H. 1997. *Econometric analysis.* 3rd edition. Prentice Hall.

Grepperud, S. 1996. Population pressure and land degradation: The case of Ethiopia. *Journal of Environmental Economics and Management* 30: 18–33.

Hagos, F. 2003. *Poverty, institution, peasant behavior, and conservation investments in Northern Ethiopia.* Department of Economics and Social Sciences, Agricultural University of Norway, Aas. (PhD dissertation, no. 2003:2)

Hailu Araya (2010). *The impact of compost in soil fertility and yield: A case of Tigray Region, Ethiopia.* Hohenheim University, Germany. (PhD dissertation)

Hailu Araya & S. Edwards, 2006. *The Tigray experience: A success story in sustainable agriculture.* Environment and Development Series No. 4. Third World Network, Penang.

Hemmat, A. & O. Taki. 2001. Grain yield of irrigated winter wheat as affected by stubble-tillage management and seeding rates in Central Iran. *Soil and Tillage Research* 63: 57–64.

Kassie, M., J. Pender, M. Yesuf, G. Köhlin, R. Bluffstone & E. Mulugeta. 2008a. Estimating returns to soil conservation adoption in the northern Ethiopian highlands. *Agricultural Economics* 38: 213–32.

Kassie, M., M. Yesuf & G. Köhlin. 2008b. *The role of production risk in sustainable land-management technology adoption in the Ethiopian highlands.* EfD Discussion Paper 08-15. Resources for the Future, Washington DC.

Kassie, M. & S.T. Holden. 2007. Sharecropping efficiency in Ethiopia: Threats of eviction and kinship. *Agricultural Economics* 37: 179-188.

Lee, D.R. 2005. Agricultural sustainability and technology adoption: Issues and policies for developing countries. *American Journal of Agricultural Economics* 87(5): 1325–34.

Mas-Colell, A., M.D. Whinston & J.R. Green. 1995. *Microeconomic theory.* Oxford University Press, New York.

Mesfine, T., G. Abebe & A.R.M. Al-Tawaha. 2005. Effect of reduced tillage and crop residue ground cover on yield and water use efficiency of sorghum (*Sorghum bicolor* [L.] Moench) under semi-arid conditions of Ethiopia. *World Journal of Agricultural Sciences* 1(2): 152–60.

Michaele Kahsay. 2005. Assessment on the economics of organic farming in Tahtay-Maichew Woreda in 'Tabia' Akab-Seat, May-Siye, May-Berazio and Kewanit. Unpublished report from 10th Cycle Practical Attachment Program of Mekelle University, Faculty of Dryland Agriculture and Natural Resource, Department of Natural Resource Economics and Management (NREM).

Neill, S.P. & D.R. Lee. 2001. Explaining the adoption and disadoption of sustainable agriculture: The case of cover crops in northern Honduras. *Economic Development and Cultural Change* 49(4): 793-820.

Place, F. & P. Dewees. 1999. Policies and incentives for the adoption of improved fallows. *Agroforestry Systems* 47: 323-43.

Sasakawa-Global 2000. 2004. *Proceedings of the Workshop on Conservation Tillage.* Melkassa Agricultural Research Center, East Shewa, Ethiopia, April 6, 2004.

Sasakawa Africa Association. 2008. Country profile: Ethiopia. Available at http://www.saa-tokyo.org/english/country/ethiopia.shtml (accessed October 2008; verified 23 March 2010).

Shiferaw, B. & S.T. Holden. 1998. Resource degradation and adoption of land conservation technologies in the Ethiopian highlands: A case study in Andit Tid, North Shewa. *Agricultural Economics* 18: 233–47.

Twarog, S. 2006. Organic agriculture: A trade and sustainable development opportunity for developing countries. *In*: *UNCTAD Trade and Environment Review 2006.* United Nations, New York and Geneva.

UNCTAD & UNEP. 2008. *Organic agriculture and food security in Africa.* UNEP-UNCTAD Capacity-Building Task Force on Trade, Environment and Development. United Nations, New York and Geneva.

ORGANIC AGRICULTURE AND FAIR TRADE IN WEST AFRICA

Olugbenga O. AdeOluwa

CONTENTS

TABLES

ABOUT THE AUTHOR

Olugbenga O. AdeOluwa is with the Department of Agronomy, Faculty of Agriculture and Forestry, University of Ibadan, Nigeria. He is also the National General Secretary, Nigerian Organic Agriculture Network (NOAN) and the International Federation of Organic Agriculture Movements (IFOAM) Contact Point Coordinator, Nigeria.

ACKNOWLEGDEMENTS

Suggestions and technical inputs from Hervé Bouagnimbeck of the IFOAM Africa Office, Bonn; Alastair Taylor, Agro Eco Louis Bolk Institute, East African regional office, Kampala, Uganda; and Cora Dankers of the Food and Agriculture Organization of the United Nations (FAO) to this document are highly appreciated.

INTRODUCTION

Organic agriculture is a production system that sustains the health of soils, ecosystems, biodiversity and people. It relies on ecological processes and nutrient cycles adapted to local conditions, rather than the use of external inputs with adverse effects. Organic agriculture combines traditional knowledge, innovation and modern science to benefit the shared environment and promote fair relationships and a good quality of life for all involved (IFOAM, 2004).

Organic agriculture is developing rapidly and is now practised in more than 120 countries of the world. As at the end of 2007, almost 32.2 million hectares of land were managed organically all over the world by 1 219 526 farmers, of which the majority (43.5 percent) is in Africa. The United Nations Conference on Trade and Development (UNCTAD) and United Nations Environment Programme (UNEP) (2008) reported that organic agriculture can increase agricultural productivity and can raise incomes with low cost, locally available and appropriate technology, without causing environmental damage. Further, the report showed evidence that organic agriculture can build up natural resources, strengthen communities and improve human capacity to cope with environmental challenges, thus improving food security by addressing many different causal factors simultaneously. The benefits of organic farming for Africa are numerous, from increasing yields and conserving water in semi-arid areas and combating desertification, to debt reduction for farmers, strengthening of social systems and maximization of environmental services.

West Africa is lagging behind in the development of organic agriculture with only ten out of the 16 countries therein having a formal record of organic agriculture activities. Overall, West Africa has a total of 733 359 000 ha of land (FAO, 2000), of which 50 568 ha is under organic agricultural production (Willer and Kilcher, 2009). This is 0.007 percent of the total land in the region and an indication of the low status of organic agriculture activities in West Africa, which mainly focus on vegetables, fruits and fibre products. In 2008, operators were private sector business people while most governmental organizations provided little or no support to the development of organic agriculture.

Apart from information sharing among members of the International Federation of Organic Agriculture Movements (IFOAM) in the region and perhaps members of the Institut Africain pour le Développement Economique et Social–Centre Africain de Formation (INADES-Formation), interaction across countries in the region among organic agriculture stakeholders is low. Even within countries, few interactive efforts are known among the operators. This situation is not conducive for the growth of the organic sector. The national organic movements known in West Africa are the Organic Movement of Mali (MOBIOM), the National Federation of Organic Producers of Senegal (FENAB), the Ghana Organic Agriculture Network (GOAN), the Nigerian Organic Agriculture Network (NOAN) and L'Organisation Béninoise pour la Promotion de l'Agriculture Biologique (Benin Organization for the Promotion of Organic Agriculture) (Willer and Kilcher, 2009).[1]

One of the steps for proper development of the organic sector in the region is to encourage cooperation among stakeholders. This action often leads to synergies that can fast track development of the organic sector. Benefits that can be derived from the influence of collaborative efforts among similar organizations cannot be overemphasized. Hence, practitioners of organic agriculture in West Africa have a lot to benefit from working together in order to catch up with global developments. Some national organic agriculture movements in West Africa have already shown interest in collaborative efforts within the region but much is yet to be done to fast track this development.

In East Africa, the cooperation among organic agriculture practitioners supported the increase in organic agricultural activities and the development of the East African Organic Products Standard. The move to develop the East African Organic Products Standard started with three countries, namely Uganda, Kenya and Tanzania, and by 2009, up to nine countries in the region were involved. This cooperation as well as other projects such as the Export Promotion of Organic

---

1   A complete list of the West African members of IFOAM can be found at the IFOAM website: www.ifoam.org

Products from Africa (EPOPA) with the involvement of many donors together with national and international actors has led to increased production and export of organic agriculture produce which is contributing to the gross domestic product (GDP) of the countries involved.

TABLE 1

West African countries with known areas of land under organic agriculture systems as of December 2007

COUNTRIES	LAND (HA)	NUMBER OF PRODUCERS
Benin	1 488	2 354
Burkina Faso	7 267	5 808
Ghana	24 449*	3 900*
Guinea Bissau	5 600	401
Ivory Coast	943	No data
Mali	3 402	7 526
Niger	131	No data
Nigeria	3 154	No data
Senegal	1 589	1 306
Togo	2 545	4 183
Total	50 568 (5.81% of that in Africa)	17 592 (3.39% of that in Africa)
Africa	870 329	529 986
World	32 221 311	1 219 526
Uganda	296 203	206 803
Tunisia	154 793	862

* 2008 data

Source: Adapted from Willer and Kilcher (2009)

CURRENT STATE OF ORGANIC AGRICULTURE IN WEST AFRICA

In spite of the low level of activities in the organic agriculture sector in West Africa, the region has some strengths that can be exploited for its accelerated development.

o **Evolution of national organic agriculture movements in the region:** West Africa is currently experiencing an increase in the number of national organic agriculture movements. In 2003, IFOAM (2003) reported little evidence of emerging or 'joined-up' agro-ecology movements in the region. Then, it was only in Senegal and Ghana that individuals were found promoting organic agriculture. However, by 2009, things were changing; national movements were emerging. Currently, at least five such movements are known in West Africa. These national organic agriculture movements are either full members of IFOAM or related in some way to IFOAM. Information flow and interaction among IFOAM members could be harnessed for a positive evolution of organic agriculture practices.

○ **Majority of farmers disfavour the use of synthetic inputs**: Traditional farmers make up to 70 percent of the farming population. They seldom use synthetic inputs in their fields. For example, farmers producing yam, cassava and most fruit tree crops do not use any chemical inputs. Thus, conversion to organic agriculture would be easy for these farmers if they are well informed about its benefits.

○ **Biodiversity**: One of the characteristics of a traditional African farm is its great biodiversity. A farmer can have up to ten or more varieties of crops and different types of domestic animals on her/his farm. This fits the principle of ecology in organic agriculture. Many benefits are found from keeping good diversity of both flora and fauna in the farm. Many problems such as poor soil fertility, pest and diseases that deter many farmers from practicing organic agriculture can easily be managed with good biodiversity. Again, agro-ecology in the region is also very diverse; West Africa has many of the agro-ecologies found in Africa.

○ **Good trade relationship between ECOWAS states**: The Economic Community of West African States (ECOWAS) encourages free trade among countries in the region. This situation provides potential opportunities to move both inputs and produce within the region for effective organic agriculture business.

○ **Increased interest of stakeholders in organic agriculture**: With the increase in awareness of the benefits from organic agriculture, more people are getting interested in organic produce. Both sides of demand and supply are increasing.

UPDATE ON ORGANIC AGRICULTURE ACTIVITIES IN WEST AFRICA
Ghana

Ghana's capital, Accra, has the country's first organic restaurant – Café Baobab – where organic menus are offered to the customers. The establishment of this restaurant aims at initiating a farm to restaurant marketing strategy, where organic producers and consumers can meet. Therefore, the growth of local markets for organic produce can be triggered. Café Baobab was formally opened in July 2007 with the help of GOAN, by Lady Tangi, an Australian whose husband is a Ghanaian.[2]

Key organic products from Ghana

Ghana's main organic export commodities are palm oil and fresh fruits. The certification of farms already using organic methods is making progress. Other key organic products include cocoa, banana, cashew nut, culinary herbs, cereals, vegetables, cotton and shea butter (IFOAM, 2003). Ghanaian non-governmental organizations (NGOs) and farmers' groups promote the expansion of organic production in the existing product range as well as in new sectors. Indigenous groups are active in developing and disseminating improved organic farming methods.

The network

GOAN is the main grouping of organic NGOs and trade associations in Ghana, which is working actively with concerned organizations such as the International Trade Centre (ITC), Henry Doubleday Research Association (HDRA), the Department for International Development (DfID) of the United Kingdom and Pesticide Action Network, United Kingdom (PAN-UK) in developing the organic sector in the country.

---

2   *Source*: First organic restaurant in Ghana, by IFOAM Africa Centre Office, 16 Jan 2008. Available at http://www.modernghana. com/news/153740/1/first-organic-restaurant-in-ghana.html (accessed 8 February 2010; verified 21 March 2010).

It has about 150 member groups and organizations, representing a thousand individual members.

With the assistance of HDRA and PAN-UK, GOAN in 1995 established an agriculture centre to provide information, training and advice on organic agriculture practices. It also has links with research institutes to examine alternative methods of pest control, particularly for cocoa, oil palm, cotton, cereals, fruits and vegetables (ITC, 2009).

Nigeria

Organic agriculture in an organized manner is still young in the country, with less than five years of experience. As of 2007, Nigeria had 3 154 hectares under organic agriculture, of which 50 ha were fully converted (Willer and Kilcher, 2009).

Practitioners are still few despite the great potential for organic agriculture. The following are the main stakeholders:

- **Dara/Eurobridge Farm**, which is the known pioneer organic farm in Nigeria that produces lemongrass, turmeric, ginger, plantain and medicinal herbs.
- **Organic Agriculture Project in Tertiary Institutions in Nigeria (OAPTIN)**, which organized a pioneering network in 2004. Its activities focus on capacity building and networking of academics in organic agriculture.
- **Olusegun Obasanjo Centre for Organic Agriculture Research and Development (OOCORD)**, which was established in 2007 and is the first of its kind in Nigeria. It focuses on research and development in organic agriculture.
- **Nigerian Organic Agriculture Network (NOAN)**, which was formed as an initiative of OOCORD and designated to be an umbrella body for organic agriculture activities in Nigeria in August 2008. Its function is to network organic agriculture organizations in Nigeria.
- **Organic Farmers Association of Nigeria**, which coordinates the activities of Nigeria's organic farmers.
- **Organic Fertilizer Association of Nigeria**, which coordinates the activities for organic agriculture fertilizer production.

o **Others** – "Nigeria Go Organic", which is currently focusing on a campaign for "Ibadan Go Organic", organic beekeepers, snail keeping, etc.

First West African Summit on Organic Agriculture

The first West African Summit on Organic Agriculture took place at the University of Agriculture, Abeokuta, Nigeria, between 17 and 21 November 2008 with the theme "Organic Agriculture and the Millennium Development Goals". The keynote address was given by Moses K. Muwanga, Chief Executive Officer of the National Organic Agriculture Movement of Uganda (NOGAMU) and Member of the World Board of IFOAM.

The idea of having a summit on organic agriculture in West Africa was raised among the West African participants at the East African Organic Conference in Dar es Salaam, Tanzania, in June 2007.

Nigeria was chosen to host the maiden edition of the summit, as it is also the headquarters of ECOWAS. The event was hosted by OAPTIN in collaboration with the University of Agriculture, Abeokuta.

The summit aimed at:
o Disseminating current information, indigenous technologies and competences in organic agriculture in West Africa;
o Providing a forum for sharing experiences among organic agriculture scientists and practitioners in West Africa; and
o Enhancing international collaboration among higher educational institutes and other stakeholders in organic agriculture in West Africa and developed countries.

The event received sponsorship for participants from IFOAM, Association of African Universities (AAU), UNCTAD, GOAN, Sustainable Agriculture Development Network (REDAD) of Benin, Agrecol-Afrique of Senegal, INADES-Formation of Cote d'Ivoire and Agro Eco of the Netherlands. There was also local funding through donations and the registration fee.

A total of 119 participants were registered including representatives from 16 countries: Benin, Cote d'Ivoire, Germany, Ghana, Kenya, Namibia, New Zealand, Nigeria, Senegal, South Africa, Sudan, The Netherlands, Uganda, United Kingdom, Zambia and Zimbabwe.

Special events at the summit included:

- Training by IFOAM on advocacy and Participatory Guarantee Systems (PGS);
- Prioritization of advocacy needs for organic agriculture in Africa;
- Experience sharing among organic agriculture practitioners in Africa;
- Networking of organic agriculture practitioners (producers, marketers and researchers) in West Africa; and
- Organic agriculture trade opportunities within West Africa.

The following resolutions were adopted:

- Participants were charged to step up the level of advocacy on organic agriculture at national, regional and continental levels;
- Participatory Guarantee Systems should be adopted to increase local organic market systems while third party certification could be used for international markets;
- A steering committee was constituted to work out a *modus operandi* for formation of a West African Network for Organic Agriculture in consultation with IFOAM;
- Next West African Summit on Organic Agriculture to be hosted by Ghana in 2010; and
- Positive spirit of the summit to be maintained and the tempo sustained at continental level.
- As a result of the summit, the organizers look forward to seeing:
- Food security improved through better availability and accessibility of organic produce;
- Improved and sustained environmental services and biodiversity;
- Good health for people, animals, the soil and the environment;
- Improved livelihoods through fair trade organic markets;

- Good spread of information on organic agriculture in research, extension, advocacy; and
- Abundant trained personnel to drive organic agriculture in the formal and informal sectors in West Africa.

ORGANIC AND FAIR TRADE EXPORTS PROJECT IN WEST AFRICA[3]

The FAO organized a project to encourage the export of organic and/or fair trade produce from Central and West Africa to Europe. The project was funded by the government of Germany. Beneficiary countries were Burkina Faso, Cameroon, Ghana, Senegal and Sierra Leone. The implementation period was September 2005 to September 2009. The aim of the project was to assist farmer groups and small exporters to help them overcome challenges and take advantage of remunerative markets.

An IFOAM survey of 2003 revealed that West Africa lagged behind other regions in the development of its organic sector. At the same time, the region has often been viewed as having the potential for developing exports of certified organic products, especially tropical fruits. Furthermore, agro-ecological initiatives promoting rural development and food security, and based on enhancing soil fertility, are a strong point in the region.

Therefore, a project was developed to support farmer groups in West Africa to build their capacity to produce and export organic and fair trade certified products in order to increase the income and food security of their farmers.

The project was conducted in a participatory manner, based on a 'tailor-made' programme of activities for each supply chain. A programme of activities, usually for one-year periods, was agreed upon in collaboration with the beneficiary group and/or exporter. In the cases of Cameroon, Burkina Faso and

---

3   Extracted from http://www.fao.org/organicag/organicexports/organicexports-home/en/ (accessed 8 February 2010; verified 21 March 2010).

the sugarloaf pineapples in Ghana, export was already taking place when the project started. In these cases, activities focused on the weakest part of the chain or on activities necessary to obtain certification. In the other cases, supply chains had to be set up from scratch and activities started from the level of organization and capabilities already present. Contracts were drawn up with the exporter and/or farmer group, in order to give them the means to implement as many activities themselves. For farmer groups, the negotiations of the contracts and meeting delivery requirements before the next instalment were learning activities in themselves and prepared them for more demanding and less flexible commercial contracts.

Product and group selection

During the project formulation phase in 2004, a market study was conducted to identify organic and fair trade growth markets in Europe. The following product categories were selected:

- Tropical fruits: pineapple and mango;
- Cocoa; and
- Shea butter.
  Subsequently, producer groups and exporters were selected based on four criteria:
- Ability to supply one of the selected products;
- Readiness to export, or already having current exports;
- Interest in organic and/or fair trade certification; and
- Possible advantages of the project for small farmers.

Results and outputs

The project has worked with seven different supply chains and more than seven farmer groups, each with their distinct characteristics and starting situation, thus outputs and results vary much according to supply chain.

Training has taken place in all farmer groups. The project trained in total:

- 2 078 farmers in organic agriculture and fair trade;
- 229 shea nut collectors in organic requirements for collection;
- 108 shea butter producers in organic requirements for butter production;
- 68 produce agents/harvesters in quality requirements and record keeping/ traceability requirements;
- 36 Internal Control System (ICS) managers/field officers/internal inspectors on their role in the ICS;
- Four documentation officers in record keeping, filing and administrative management;
- 16 executive and board members of farmers' associations in the running of their organizations, including development of sales to exporters (one group) or in direct exports (two groups); and
- Five managers of exporting farmers' organizations and one exporter in the development of their export business.

All groups were certified as planned. In terms of export development, results have been as follows:

Burkina Faso:

- BurkiNature increased exports of organic and organic/fair trade mangoes by 40 percent from 2005 to 2006, and between 2006 and 2008 by another 50 percent.
- Club des Productrices de Beurre de Karité Biologique (CPBKB) increased exports of organic shea butter five-fold in the course of the project.

Cameroon:

- Pineapple growers' association Groupement d'Interêt Commun – Union des AgroPasteurs du Cameroun (GIC UNAPAC) increased exports of pineapples by 40 percent from 2005 to 2008.

Ghana:

- Tropical fruit exporter WAD Ltd. increased sales of dried and fresh pineapple, and now buys 2.5 times more pineapples from the farmers than at the start of the project (increase of 170 percent).

- Volta Organic Mango Farmers Association (VOMAGA) started selling mangoes to processors.

Sierra Leone:

- Cocoa association Kpeya Agriculture Enterprise (KAE) exported its first container of fair trade certified cocoa in January 2009.

Impacts

The most important criterion for the success of a project is the impact on the target group. To get a better idea of the impact at farmer level, the project conducted impact surveys in all countries. With the exception of Senegal, these surveys were conducted in the period April–June 2008. The impact of later project activities and developments has thus not yet been assessed.

The impact surveys concluded that the new organic production methods have resulted in improved quality of the products. The majority of respondents also observed an increase in production, which was due to a combination of higher yields and increases in cultivated areas or in the case of shea butter, an increase in collection efforts of shea nuts and subsequent increased transformation of nuts into shea butter.

According to the organizers of this project, whether this increase in production and exports have resulted in reduced poverty and food insecurity is more difficult to ascertain for two reasons. Firstly, the impact of the adoption of the new agricultural and processing methods on the total costs of production varies considerably from one sub-project to the other. Secondly, the starting situation of each sub-project varied considerably as well as the level of poverty and food insecurity of the group members.

Concerning the cost of production, it is clear that the implementation of the organic methods generally resulted in an increase in labour costs and a decrease in the costs related to the purchasing of agrochemicals. Group marketing reduced the transportation costs of the products to the market. Regarding the variations in the living conditions at the start of the project, it can generally be concluded that the poorer the producers, the more the project's impact manifested itself in terms of poverty alleviation and food security.

In general, the project has resulted in an increase in the incomes of its participants as a result of the increase in the production volumes or the price paid to the producers. The additional income generated through the sale of certified products is mainly used for purchasing food or clothing, for paying school fees and for medical expenditures, thereby improving the living conditions and the food security of the participants. Five out of the seven sub-projects had led to the marketing of certified products at the time of the impact surveys. The producers in these groups confirmed nearly unanimously the positive impact of the marketing of the certified products by the producer groups; no disadvantages were mentioned. The impact surveys also confirmed the project's impact on employment through the creation of jobs for workers directly involved in the production of certified products, as well as for workers and administrative staff involved in production supporting services.

CHALLENGES FACING ORGANIC AGRICULTURAL DEVELOPMENT IN WEST AFRICA

Based on observation as well as a review of past studies, especially by IFOAM (IFOAM, 2004), the main challenges facing the development of organic agriculture in West Africa are briefly described below.

o **Poor local marketing**: Like any other business, organic agriculture thrives with efficient marketing systems. Generally, many farmers have been discouraged from going into organic production because of the lack of or poorly developed local markets for organic products. The majority of the organic agriculture operators in West Africa focus on export markets. Many farmers who cannot afford the cost of third party certification find it difficult to sell their organic produce at prices that would have compensated for extra efforts put into an organic system of production. However, there are exceptions; for example, Ghana and Senegal have some organic market points. In some situations, organic products are sold without any differentiation from the conventional agriculture products. A certified organic farm in Nigeria, for example, currently sells its organic lemongrass tea, turmeric and other produce to the local market, a situation many regard as under-maximization of the premium benefits in organic agriculture. However, local markets make it easy for resource-poor farmers to sell their products and interact with processors and consumers. In Europe, America and some parts of East Africa, there are organized sales outlets where consumers can get organic products. This situation has a way of encouraging producers in such regions to stay in business.

○ **Low level of organic certification**: Third party certification in Africa is mostly used as a means of accessing foreign markets. However, the majority of the farmers in West Africa cannot afford the costs of getting their farms certified through the third party system. Hence, the number of certified farms in the region is very low compared to East and South Africa. Perhaps this is due to poor public awareness of business opportunities in organic agriculture, because people in the region are not so poor to the level of finding it difficult to invest in organic agriculture. Many of the business communities in the region are oblivious to the big international market in organic agriculture. Without building awareness of the opportunities, little or nothing can be done to improve investments in the sector. However, use of a participatory guarantee system, mainly for local markets, could be explored in the region, as this does not incur such high costs as third party certification. It also has the advantage of building confidence between local producers and consumers.

○ **Little information on organic agriculture activities**: Information on organic agricultural activities in the region is meagre. Perhaps, more farmers would have become interested in organic agriculture if they had access to information showing how feasible it is to invest in the practice. Where organic agriculture is practised, producers claim increased income due to the premium price paid for their produce as well as the avoidance of costs for external inputs. The profitability of organic agriculture has been reported to attract the interest of many operators. In East Africa, the number of organic farmers has multiplied three-fold within a few years. There is no doubt that concerted information dissemination on organic agriculture in West Africa could contribute to the growth of the organic sector in the region.

- **Little or no policies to safeguard organic agriculture activities**: Most national governments in West Africa have no policies in place to safeguard organic agriculture practices in their countries. Some of the countries in the region could have some plans, but these are yet to be translated into working documents that organic agriculture producers could refer to for sustainable and confident investment in the organic agriculture sector of the region. Indiscriminate use of agrochemicals and other synthetic inputs is not yet adequately recognized as a serious problem in the region. These situations are challenging the development of organic agriculture.

- **Sourcing of appropriate inputs for organic agriculture**: A major constraint to the adoption of organic agriculture in West Africa is the lack of appropriate inputs, such as bio-fertilizers and bio-pesticides needed in organic agriculture production. It is easier to get the conventional agriculture inputs. Information on appropriate cultural methods to address the challenges of soil fertility decline and pests and diseases is not widely available to the farmers. It is a fact that producers are not able to produce beyond resources at their disposal; even if they are willing.

- **Poor private sector involvement**: The current level of involvement of the private sector in organic agriculture in West Africa is very low. This is true when one compares the situation with what is found in East, Northern and Southern Africa. In other places of the world, private firms invest heavily in production, processing and marketing of organic agriculture products. The resultant effects are increased activities and the expansion of the organic agriculture sector.

- **Lack of technical assistance**: Compared to East and Northern Africa where there are a reasonable number of organic agriculture service providers, the opposite is the situation in West Africa. Often, farmers need technical assistance in terms of soil fertility maintenance, pest and disease control, mechanization and acquisition of useful skills that can enhance good production in the organic sector.

o **Low interaction among national organic networks in the region**: Interaction among national organic networks as well as other stakeholders in West Africa is still very low. Though many steps are on-going to bridge gaps among operators, much still needs to be done. The West African Organic Agriculture Summit, reported above, was held to start to change the situation, but the fact is that a lot of efforts are still needed to bring stakeholders together.

o **Lack of funds to foster organic agriculture projects**: Unlike projects in conventional agriculture, organic agriculture projects receive less attention from most donors. Much still needs to be done in terms of proper awareness and public education on the benefits of organic agriculture. Major operators in the organic system in the region are private farms that find it difficult to spend their resources on public enlightenment.

FAST TRACKING DEVELOPMENT OF ORGANIC AGRICULTURE IN WEST AFRICA

The following interventions are needed for proper development of organic agriculture activities in West Africa:

Research, training and development

1. Carrying out of a comprehensive survey of the current situation in the field in order to determine the advantages and constraints to the development of organic farming in West Africa, because little or no information is available on specific issues.
2. Land capability and suitability evaluation of areas that could be adapted to organic farming within the different agro-ecologies in the region.
3. Development of support measures for organic farming. There is a need to develop collaborative training modules for organic farming to build capacity of organic

agriculture operators in the region. Establishment of technical support structures, workshops and public awareness on organic agriculture focusing on benefits derivable from it as well as mass sensitization through posters, newsletters, etc.

Administration and policy

1. There should be strong advocacy and lobbying to promote organic agriculture in the region.
2. Efforts should be made to make governments of the region come up with national land tenure policies to encourage organic agriculture.
3. There is a need for integration of organic farming into the agricultural policy of ECOWAS states.
4. Governmental and private investments in organic resource inputs should be encouraged.
5. Rationalization of organic farm certification costs through the establishment of local certification bodies and encouragement of more private investment in this aspect of organic agriculture.
6. Organization of farmers' associations along commodity lines e.g. rice, cocoa, cassava growers' associations.
7. Strengthening of national organic agriculture movements through regular meetings to encourage more farmers to go organic.
8. Regular inventory of organic farms and farming activities in the region should be done at country level.
9. Establishment of ECOWAS organic produce standard could encourage inter-country organic business activities in the region.
10. There should be solution centres for organic agriculture in the form of 'Help Desks' (international, national and regional) in the region.
11. There should be regional and national organic agriculture coordinating centres.
12. Model pilot projects on organic agriculture should be instituted in strategic locations in countries within the region to serve as learning places for others.

Marketing and quality control

1. Potential producers and consumers of organic products should be encouraged to invest in the organic sector.
2. There should be the organization of standard sales outlets for organic products to bring organic produce closer to potential buyers.
3. Regional regulations and quality control measures should be developed.
4. Cooperation with appropriate international bodies, such as ITC, UNCTAD, UNEP and IFOAM should be encouraged in the region.

CONCLUSION

In 2009, West Africa still had a low level of organic agriculture activities. However, considering the potentials within the region that can support organic agriculture, 'the sky is the limit' for the expansion of organic agriculture. Though there seem to be a lot of problems facing the development of organic agriculture, all these are surmountable if most of the necessary interventions suggested above are put in place. The fact that West Africa has a diversity of agricultural ecological zones, as well as other strengths, could provide a comparative advantage for the region over others in Africa for diversified organic agriculture production.

REFERENCES

FAO. 2000. *Global forest resources assessment 2000.* FAO, Rome. Available at http://www.fao.org/DOCREP/004/Y1997E/y1997e0j.htm (accessed 27 August 2009; verified 21 March 2010).

IFOAM. 2003. *Organic and like-minded movements in Africa.* IFOAM, Bonn.

IFOAM. 2004. Network building for lobbying in Africa. Compiled by Souleymane Bassoum, René Tokannou and Nguji Mutura. IFOAM, Bonn.

ITC. 2009. Country profile: Ghana. Organic Link. ITC, Geneva. Available at http://www.intracen.org/organics/Country-Profile-Ghana.htm (accessed 28 August 2009; verified 21 March 2010).

UNCTAD & UNEP. 2008. *Organic agriculture and food security in Africa.* UNEP-UNCTAD Capacity-building Task Force on Trade, Environment and Development. United Nations, New York and Geneva.

Willer, H. & Kilcher, L. (eds.) 2009. *The world of organic agriculture – Statistics and emerging trends 2009.* IFOAM, Bonn, FiBL, Frick and ITC, Geneva.

ORGANIC TRADE PROMOTION IN UGANDA: A CASE STUDY OF THE EPOPA PROJECT

Alastair Taylor

CONTENTS

ABOUT THE AUTHOR

Alastair Taylor is the Regional Manager, Agro Eco Louis Bolk Institute – Eastern Africa Office (AELBI) (www.louisbolk.org). He was the former Country Manager of Export Promotion of Organic Product from Africa (EPOPA) Uganda.

INTRODUCTION

The Export Promotion of Organic Products from Africa (EPOPA) project was conceived by the Swedish International Development Agency (Sida) as a result of them commissioning a consultancy to explore opportunities for African countries to utilize their low input agriculture for commercial advantage. This exploration took place in the early 1990s and by 1995, the first phase of EPOPA was tendered for implementation in Uganda, where it was felt that the production, market and policy opportunities were best suited for successful implementation. The initial tender was awarded to the Dutch organic consultancy company, Agro Eco, which has now merged with the Louis Bolk Research Institute to become Agro Eco Louis Bolk Institute (AELBI). In the mid-1990s the organic market was growing, but its opportunities were still not fully realized by many traders operating in Africa.

The first work of the EPOPA consultants was to present the opportunities of the organic market place to existing export companies operating in Uganda. These were mainly dealing in commodity crops and hence this was where the interest was born, firstly in the opportunity to export organic cotton and secondly in organic coffee, mainly Arabica. EPOPA was conceived as a private sector programme and the entry point for any intervention was close co-operation with an existing export company and the identification of production areas where organic production would face the least risk. The market focus had to be on certified organic products and the production base had to be smallholder farmers.

An existing company, an organic focus and smallholder production were the key ingredients from the company side and then to these were added quality technical support and financial injections from the EPOPA side. Each project was considered as an investment by EPOPA into the company and hence the company had to be open with EPOPA about its circumstances, and be willing to share costs in the development of the organic product line. The success of the intervention was partly judged by the 'pay-back' period that was required for the EPOPA investment to be recouped from increased incomes to the supplying farmers – mainly considering the organic premium.

KEY INGREDIENTS FOR SUSTAINABILITY

It is worth looking at some of these ingredients in some more detail and considering their bearing on the eventual market success of the project, which actually reflects its sustainability.

Existing company with export experience

Export-orientated marketing is not easy. Seller and buyer are far from each other, transport chains can be unreliable, paperwork can be daunting and the time lag between purchasing from farmers and payment by the buyer can be very long. Although EPOPA was able to offer technical support in some of these areas, especially where they were unique for organic trade, it was not its major activity area. Trading competence was assessed during the company selection process and, although there were exceptions to this guideline, the best results were generally achieved when previous export experience was present.

Organic products

EPOPA wanted to get extra money to the farmers and thus help them relieve their poverty situation. Much production in Uganda is 'traditional' and uses very few, if any, chemical inputs. But organic is not simply about not using chemical inputs; it proactively cares for the soil, the plant or animal, the environment within which production is taking place and the people who are involved in production. Such priorities are summed up in the International Federation of Organic Agricultural Movements (IFOAM) principles of "Health, Ecology, Fairness and Care".

With the EPOPA focus on organic production and marketing there was the opportunity to apply these principles and increase the quality and quantity of production. Often these improvements were the first benefits that farmers realized by entering the organic project. Naturally, the key market opportunity was normally seen as the premium price and increased market opportunity through being able to promote and sell certified organic products, but the premium varies and an organic crop enters a competitive market similar to any other. Through better soil fertility management, soil and water conservation practices, and the application of other good agricultural practices, organic coffee farmers were able to increase the grade quality of their coffee and quantity available for the market.

The organic premium was targeted to be 20 percent above the conventional price, but was seen to vary from zero percent when the local demand for a product was very high, as had been the case with sesame in 2009 due to the market demand from Sudan and Kenya, to over 200 percent when the organic opportunity was unique and demand high, such as the case with a number of organic fresh fruits. In 2006/2007, 54 000 farmers associated with the EPOPA programme sold over USD 12.6 million; the extra income – or organic premium – they realized was in the region of USD 2.6 million and the premium for the exporters was over USD 5 million.

Smallholder production

It has been stated by Ugandan policy makers that the predominance of smallholder farmers is a negative factor holding back the country's development potential. The impact of EPOPA and the successful organic trade that has been the result of its implementation has shown this not to be the case.

Organic certification within the EPOPA programme was through "smallholder group certification". Each smallholder farmer is given a unique code, signs a contract with the buyer, is visited at least once a year by an internal organic inspector – who is also able to offer the farmer extension advice on organic practices – and the farmers feel a pride in knowing that they are internationally certified as organic. This organization is described in a document called an "Internal Control System" (ICS) and the discipline that surrounds being part of an ICS has also been seen to boost the farmers' ability to plan their farms in a better way and even take better care of their families.

Under the influence of the EPOPA project alone within Uganda, from 2004 to 2008, there were over 87 000 smallholder farmers certified as organic providing cocoa, coffee, cotton, tropical fruits, hibiscus, shea, sesame, chillies, vanilla, lemongrass, cardamom and other crop products to their contracted buyers. The value of these products in 2006/2007 on the export market was over USD 25 million.

Quality technical support

About 50 percent of the money supplied by Sida towards the EPOPA programme went towards providing technical support along the organic value chain from production, through certification and processing to end marketing. As far as the partner companies were concerned this was seen as free and expected support from the EPOPA consultant team, but when a value is put on this its importance can soon be realized.

The process of smallholder group certification is not very complicated, but it does require careful attention to detail if certification is going to be readily achieved. Normally, organic production of tropical products is not that difficult, but when pest, disease and nutritional problems arise it was a great relief for the farmers to know that EPOPA technical support was just a phone call away and when export paperwork and market presentation were causing concern, again EPOPA technical advice was available to assist the company through the issues. When promoting a new system or market opportunity, it is very important that the process should not fail, and that risks to companies and farmers should be minimized. Hence, this is where quality technical support is essential.

Financial injection

EPOPA did not give loans and would not normally give grants, but it did assist in covering a number of costs related to the organic conversion process, including direct support towards the cost of certification for the first two years of the three-year project period. This financial support helped remove some of the risk felt by companies as they embarked on organic marketing – new products in a new market with uncertain expectations.

A major EPOPA cash injection was to support the selected companies to directly contact potential buyers, mainly through participation in trade fairs, such as Biofach Germany, which is the premier global organic trade fair, but also presenting products of the selected companies whenever the chance arose through EPOPA consultants based in Europe. In fact – EPOPA had a whole section that was focused on market presentation and promotion – these consultants continue to play this linking role despite the EPOPA programme having come to an end at the end of 2008.

Business approach

It was at times difficult for some in Sida to come to terms with dealing with large companies that have significant resources and considering donor support as a company investment, but this was appreciated in the long run when results showed significant returns on Sida investment. In the second phase of EPOPA – from 2002 to 2008 – the Sida investment was about USD 5.5 million and yet the extra annual income received by farmers in 2006/2007 was about USD 2.6 million, which meant that the Sida investment was recovered in just two years.

With this commercial foundation, business does not come to an end when the donor support stops. This is ably demonstrated by the fact that, in 2009, at least 14 of the 19 companies assisted by EPOPA continued to be involved in organic trade, and, global recession excepted, will continue to do so for the foreseeable future. Thus, the premium income will continue to come to their contracted farmers.

EXPANDING THE APPROACH

In 2001, the first phase of EPOPA was evaluated and a second phase was put out to tender and won by a joint tender of AELBI and the Swedish organic consultancy company, Grolink. The basic format of "EPOPA 2" was the same and the six foundation stones listed above remained. However, in other ways the approach was expanded – non-traditional crops were added to the commodity crops and the organic sector in Uganda, within which the export trade taking place, was also developed, i.e. the organic movement, organic certification services, capacity building, policy advocacy, local/regional marketing and value addition.

By the close of EPOPA in 2008, 19 companies had benefited directly from EPOPA support and about 15 products were available in fully certified organic form, UgoCert Ltd was fully accredited to offer certification services, Uganda Martyrs University had adopted the EPOPA-commissioned organic sector training as an official option within its short course programme and the Ugandan National Organic Policy was in its final stages of government approval.

EPOPA operated in Uganda in two phases between 1995 and 2008, a long period of commitment from Sida and perhaps one of the reasons for the quality impact – there was time to develop, learn from mistakes and adjust for better impact. Sida was able to measure the impact of EPOPA through the additional income received by farmers, but many others measured the success of a trade-based programme through the success of marketing and the growth of the companies operating in these markets. Some of the marketing opportunities and challenges are discussed below.

MARKETING OPPORTUNITIES

- **Niche market**: Organic certification allowed the products to enter a niche market, which was growing rapidly on a global scale. Quality products could be more easily recognised and the reputation of Uganda as a supplier of organic tropical products is now recognised internationally.
- **Product quality**: Product quality is both an opportunity and a challenge. The opportunity is that Uganda has the potential to supply the best quality products in terms of active content e.g. vanillin in vanilla, sweetness in tropical fruits.
- **Production cost**: With production coming from smallholder farmers, the naturally good growing environment and the low incidence of pests and diseases, the

cost of production is low and even with the high cost of transport from Uganda, product costs can remain competitive.

o **Strong organic movement:** The National Organic Agricultural Movement of Uganda (NOGAMU) is now a well-established and recognised membership organization for those involved in the organic sector. Its officers have managed to raise funds to support the organic market chain, including capacity building support to help companies meet export requirements and to promote Ugandan products within the global organic market.

o **Marketing support:** A number of trade-based support initiatives are seeking to assist Ugandan companies meet international marketing requirements, such as the EU-Pesticides Initiative programme, the Dutch Centre for the Promotion of Imports from Developing Countries (CBI) market assistance programme, and the Business to Business (B2B) programme of Danida. These programmes focus on all exports, but companies trading in organic products are able to benefit from their support.

o **Contract farming arrangement:** An agreement between the farmer and the buyer is one of the key elements of group certification and this has enabled farmers to grow products knowing that they have an assured and premium-priced market to reward their efforts. The trading arrangement normally also cuts out the middle person and so more of the buyers offer prices that are able to reach the farmer.

MARKETING CHALLENGES

o **Market communication:** EPOPA dealt with Ugandan companies that were part of international trading groups that reached down to Ugandan family business entrepreneurs. The marketing needs of the companies varied a great deal and often in the case of national entrepreneurs, market communication was a key learning element. The way to do business with an American, a German or a Japanese buyer is quite different and a successful manner of approach to one can often destroy a relationship with another. This could be in a face-to-face meeting at a trade show or through an email communication. The marketers had to learn how to communicate appropriately.

- **Transport:** Uganda is a land-locked country and along with the rest of the export business, this is a challenge to the organic sector. Due to the premium price of some products, airfreight became a viable option, but with increasing concerns about global warming there has come a growing concern about the environmental impact of such trading practices. For instance, the transport of fresh fruits from Uganda by container is not a viable option due to shippers not being able to guarantee delivery times and, in any case, once this option is taken the "picked-ripe" marketing slant cannot be used for market advantage and competition from West Africa and Latin America grows.

- **Market acceptance:** Some products have "known" origins and despite the excellent quality of the Ugandan products they have to battle to gain acceptance over the "normal supply location" e.g. vanilla is from Madagascar, fresh pineapples are from Ghana, African Arabica coffee is from Ethiopia or Kenya.

- **Competition:** Organic status does not guarantee a market and all the market norms have to be played if products are to be marketed successfully. Some potential entrepreneurs unfortunately imagine that organic certification is the magic panacea to their marketing problems; in reality they were probably poor marketers before they became organic.

- **Product specification:** Global markets are very demanding on the specifications they require for purchased products – grain cleanliness should be 99.9 percent, fresh pineapples should be no more than 1.5 kg per fruit, cocoa should be properly fermented. All these can be achieved in Uganda with careful production techniques, grading and processing investment, but sometimes the samples are produced to meet the specification, however, as time goes by, the quality controls lapse, specifications are not met and the market opportunity closes.

- **Product quality:** This is related to the specification – international markets operate to international standards and organic status is not an excuse for poor quality in other areas. For some crops, and especially for processed products, this quality might need to be officially monitored through the application of additional certification procedures such as Hazard Analysis and Critical Control Points (HACCP), Global Gap or Utz (Mayan for "Good") Certification. This requires additional expertise, investment and time, but cannot be avoided if marketing is going to be successful.

- **Product quantity:** Apart from the major commodity crops, Ugandan production, especially when local entrepreneurs are involved, is still on a small scale, for example, a company producing 500 kg of dried fruit per month. Direct retail marketing is a means of achieving a market for such quantities and this has worked well for some exporters involved in vanilla and dried fruit marketing, but the quantities able to be marketed through such routes are limited. For a number of products, companies could benefit by combining production and marketing jointly and yet the confidence to have such an arrangement is limited.

- **Value addition demands:** Value addition is being promoted by the government of Uganda and seems to be a natural opportunity, but processing requires investment in machinery, careful consideration of the processing demands, availability of appropriate packaging, application of quality standards and the fact that many developed countries want to do the processing themselves means that import incentives only relate to raw materials. Packaging demands can also change rapidly: one EPOPA-supported company in Tanzania started with jars, then changed to sealed cans and then to ring pull cans to meet the market demand.

- **Government support:** The Ugandan government has allowed free trade and there are no taxes on exports, but, at the same time, their support for the promotion of the organic market has been limited and much had to be borne on the shoulders of EPOPA and now NOGAMU.

- **Strength of alternative markets:** A number of the organic export products being promoted are food products e.g. sesame, honey and fruits, and at times the local market demand can be stronger that the export demand, such as the example of sesame given above. Despite the contract arrangement, farmer allegiance is often weak and when a better market opportunity comes along they will sell into this market and the organic exporter runs out of supplies.

CONCLUSION

In conclusion, attention is drawn to an initiative that took off in the final year, 2008, of EPOPA. This was a coming together of some of the EPOPA-supported export companies to form the "Association of African Organic Exporters" (AFROEX). There are now 12 members of this association spread around East Africa and they have registered a joint brand, "Jambo Africa", and have already embarked on a joint marketing effort, sourcing of packaging and lobbying. This seems to be a very positive development as we consider the marketing challenges listed above and may be a light for the future as EPOPA has come to an end. For more information on AFROEX refer to www.organicjambo.com.

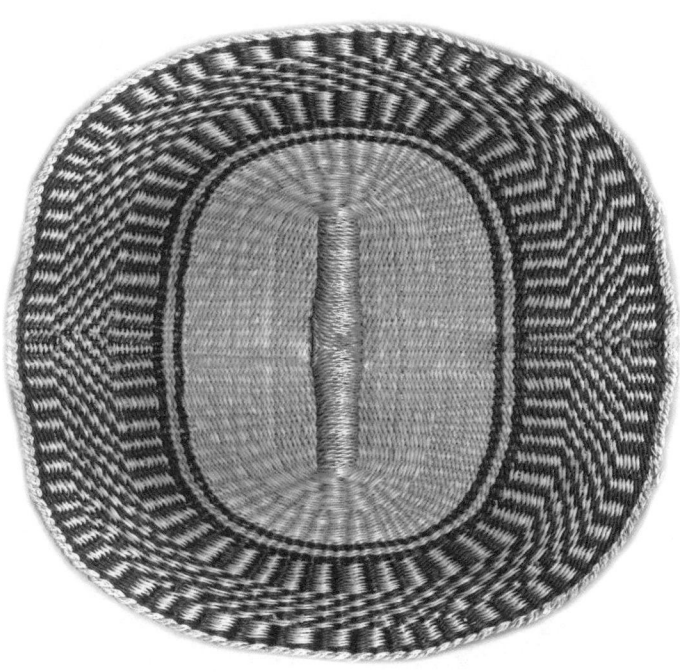

ESTABLISHING A COMMUNITY
SEED SUPPLY SYSTEM:
COMMUNITY SEED BANK
COMPLEXES IN AFRICA

Melaku Worede

CONTENTS

FIGURES

ABOUT THE AUTHOR

Melaku Worede is senior advisor to the non-governmental organizations (NGOs),
Seeds of Survival/International and Unitarian Service Community of Canada. He
is a plant breeder who worked for many years as the General Manager of Ethiopia's
Plant Genetic Resources Centre, now the Institute of Biodiversity Conservation,
establishing the first community seed bank for restoring, conserving and enhancing
Ethiopian farmers' varieties with local communities.

DIVERSITY: A KEY FACTOR FOR SUSTAINABLE PRODUCTIVITY

The pool of genetic variability within a species allows it to adapt to environmental changes. This has a special significance for the maintenance and enhancement of productivity in agricultural crops in a region such as Sub-Saharan Africa where the agro-climatic conditions are very varied resulting in diverse ecosystems for farmers to grow their crops. Such diversity provides security for the farmer against diseases, pests, drought and other stresses. It also allows farmers to exploit the full range of the region's highly varied micro environments differing in characteristics such as soil types, water availability, temperature, altitude, slope, and overall fertility.

Diversity among and within species is especially significant to Africa's farmers as it represents an important resource for farming communities to meet their subsistence needs. A wide variety of plant and animal species provides material for food, fibre, medicine, and other socioeconomic uses. Such diversity is crucial to sustain current production systems, improve peoples' diets and maintain life support systems essential for the livelihoods of local communities.

Maintenance of diversity both within and among species is, therefore, crucial to supporting and developing agriculture that is ecologically sustainable and helps local communities cope with the challenges of climate change. This is especially true for smallholder farmers practicing agriculture under low-input conditions on difficult, often degraded lands.

THE THREAT OF GENETIC EROSION

The broad range of genetic diversity existing in Africa, particularly in traditional and wild gene pools, is presently subject to serious genetic erosion and irreversible losses. This threat, which involves the interaction of several factors, is progressing at an alarming rate. The most crucial ones include displacement of indigenous farmers' varieties (landraces)[1] by new, genetically uniform crop cultivars, changes in agricultural development strategies and systems and/or land use, destruction of habitats and ecosystems, and reduction in rainfall leading to drought.

The drought that prevails in many parts of the African continent has directly and indirectly caused considerable genetic erosion, and at times has even resulted in massive deaths among people, their domestic animals and plants. The famine that persisted in some parts of Ethiopia in the last two decades, for example, has forced farmers to eat their own seed to survive or to sell their seed as a food commodity. This has often resulted in massive displacement of native seed stock by exotic seeds provided by relief agencies in the form of food grains.

THE NEED FOR RESEARCH TO CONSERVE AND ENHANCE *IN SITU* DIVERSITY

In situ (on-site) conservation of farmers' varieties on smallholder farms is providing a valuable option for conserving crop diversity (Melaku Worede, 1991). More importantly, it helps sustain evolutionary systems that are responsible for the generation of genetic variability. This is especially significant in the many parts of Africa subject to drought and other stresses, because it is under such environmental extremes that variations useful for stress-resistance breeding are generated. In the case of diseases or pests, this allows for continuing host-parasite co-evolution.

Also under these conditions, access to a wide diversity in local seeds probably provides the only reliable source of planting material. The ability of such materials to survive under these stresses is conditioned by their inherent broad genetic base. This is often not the case with the more uniform, new or improved cultivars which, despite their high yield potential, are less stable and not as reliable as sources of seed under the adverse growing conditions generally present in many of the drought-prone regions of Africa.

---

1 Farmers' varieties, often referred to as landraces, are crop populations that have been adapted through years of selection and innovation by farmers, their local communities and the environment to meet the conditions under which they are cultivated.

In these situations, establishment of species adapted to extreme environments in field gene banks, including semi-arid conditions, at strategic sites can provide a seed reserve for post-drought planting in places where traditional crops may have completely failed. Germplasm materials maintained in such fields could be distributed to rural farming communities, scientific institutions and others for further investigation of their potential use in plant breeding programs to improve food security.

Programmes for the evaluation and enhancement of farmers' varieties are certainly needed to stimulate the utilization of germplasm resources that are already adapted to these conditions. Also, under such extreme environments, locally adapted farmers' varieties can provide suitable base materials for institutional crop improvement programs in modern agricultural research organizations. There is, therefore, a pressing need to maintain farmers' varieties being grown under these dynamic conditions, and this is probably best achieved through farm and/or community-based conservation programs.

There are several programs in Africa that promote important community level seed activities, which have a tremendous experience and pool of expertise that should be drawn upon. The Seeds of Survival/International (SOS/I) programme initially developed in Ethiopia and supported by USC (Unitarian Service Community) Canada, for example, has done a significant amount of work in building the technical capacity of African farmers and researchers in on-farm conservation, enhancement of farmers' varieties and in community level seed production strategies. This programme still operates in countries of Eastern and Western Africa.

The Ethio-Organic Seed Action (EOSA) programme was developed from this earlier work. EOSA is an NGO promoting integrated conservation, use and management of agricultural biodiversity. With the guiding principle of "conservation through use", the programme works with community groups, government researchers, other NGOs and industry to promote greater integration, and especially the integration of producers with the market. The programme operates at local, regional and national levels and aims to help develop mechanisms to support the ability of smallholder farmers' to manage their own resource-base through community-based seed networks, building linkages between farmers and industry through local markets and the promotion of organic agriculture. EOSA has been successful in promoting agricultural biodiversity conservation and increasing the diversity of durum wheat and other field crops in the programme areas (Anonymous, 2009).

EOSA has documented reliable experiences of effectively working on the conservation and improvement of local farmers' varieties, on community seed banking systems and seed multiplication that increase options for planting materials for farmers. It is necessary to further expand and promote such experiences through networking where a regional level of exchange of experience and expertise is possible.

LOCAL COMMUNITIES AND MAINTENANCE OF GENETIC RESOURCES

Farming communities have always implemented conservation methods known to the formal sector as *ex situ* (off-field) and *in situ* (in-field) conservation strategies. They have been preserving or conserving their local crop types and varieties in gardens, back yards, fields and in their traditional storage facilities. The farm household includes small stores (clay pots, gourds, underground pits, etc.) that represent a "*de facto*" *ex situ* conservation system that is probably more dynamic than the conventional one at a formal gene bank.

Traditional agro-ecosystems are sources of expertise for a sustainable, diversity-based agriculture. Many species little known to science or industrial technology are still being managed by local communities; these all together form a complex of dynamic communal gene bank systems. Endangered plant species as well as economically and ecologically useful crop types are usually included in the system as part of the community-managed environmental protection and species conservation schemes. Such species may include various wild trees, shrubs and grasses of traditional use to the communities as food, feed, medicine and sources of materials for fuel and construction.

Rural women are key members of the society in the strategy for farmer-based genetic resources conservation. They are traditionally involved in making seed selection, cleaning, storage and utilization. They are mainly responsible for the safe storage at the household level of planting materials desired for the next season.

Seed is planted in the fields, i.e. on the same farm or in neighbouring areas where it acquired its distinctive features; it is also frequently exchanged among farmers and communities to be planted across regions differing in agro-ecological conditions. This can account for the broad range of adaptability (plasticity) inherent in such material (Melaku Worede, 1988). In places like Ethiopia, individual and communal underground seed storage is also a common practice when there is a drought leading to a famine crisis. Threatened households and communities, by tradition, bury large quantities of seed before they are forced to migrate elsewhere and then reclaim this seed for planting later when the drought crisis is over, usually within a three-year period (Melaku Worede, 1993; 1997). Maintenance of genetic diversity in this way provides a wide range of options for self-reliance in food crop production and security, thereby lowering the risks of food shortage.

Farmers in many parts of Africa also traditionally intercrop varieties and species. Thus, new variations are often created as a result of crossings within the mixtures. This has always given them the option to widen the diversity in their crops and to adopt the newly formed variations useful for sustaining productivity and other requirements through changing conditions, including climate change. The knowledge and the diversity local farmers have created have served as a basis for modern plant breeding and agricultural development (Emmanuel *et al.*, 1999).

Through continuous use and evaluation they maintain the genetic value of their varieties. This varietal genetic potential includes resisting environmental stresses, pests and diseases, as well as qualities such as palatability and storability that are well understood by the farming communities, particularly by women farmers. Special names that reflect the behaviour of those genes are usually given to these varieties. The high-lysine (an amino acid/protein deficient in most cereals) sorghum cultivar popularly known as 'wotet begunche' (milk in my mouth) in Wello, Ethiopia, is but one example.

Similarly, varieties resistant to birds, pests and microbes are given names, which either indicate the mechanisms of the resistance or special varietal behaviour that is responsible for that specific type of resistance. This indicates that traditional agricultural knowledge can serve as an important source of information in the improvement of agricultural productivity (Regassa Feyissa, 2000).

All modern forms of crop breeding are, in fact, for the most part, dependent on the diversity promoted and maintained by local farming communities.

SECURING A COMMUNITY-BASED SEED SUPPLY SYSTEM

The seed system used in most traditional farming systems is based on the local production of seeds by the farmers themselves. Farmers consistently retain seed as a security measure to provide a back-up in case of crop failures. They always store seeds for three main purposes:

- consumption;
- sale; and
- seed stock (for sowing in the next season).

Farmers practice seed selection, production, and saving for informal distribution of planting materials within and among the farming communities. Seed production in most cases is non-specialized; it is an integrated production of field crops, roots and tubers for consumption and marketing. This traditional seed supply system is an important backup to overall agricultural crop production in a country. It is mainly based on the farmers' varieties with the exception of cases where the seed system depends on improved or introduced crop varieties. Usually, dependency on introduced varieties is created by the displacement of farmers' own varieties. This is the case in many parts of Africa that have been influenced by modern commercial crop production systems. The potential use of formal seed, which is characterized by a vertically organized production and distribution of tested seed and approved varieties, has limited adaptability under the prevailing conditions resulting from climate change.

Variety use and development, seed production and storage by farmers under local conditions, and seed exchange mechanisms still remain the important components of the dynamic system that forms the most important source of food crops for smallholder farmers.

Unfortunately, the economic value given to modern agricultural crop productivity has, for the most part, neglected the important contributions made by traditional crop improvement and seed supply systems. It has also largely ignored the steady depletion of traditional crop varieties, and has become a cause for a shortage or disappearance of locally adapted seeds.

The objective of the formal seed system in most cases remains at odds with the needs of smallholder farmers, who require multiple varieties of seed for all crops, and in small amounts, at the right time and at a reasonable cost (Regassa Feyissa, 2000). Similarly, most public and private seed enterprises do not produce and distribute seeds to meet the subsistence needs of rural households or for farmers living in economically marginal and environmentally challenging areas. The private seed companies see links with such farmers as economically unviable. Therefore, in order to ensure seed security in areas where the formal seed system is ineffective in particular, the capacity of the informal seed sector should be improved for a reliable supply of locally adapted varieties.

With the advent of the modernization of agriculture and centralization of seed supply systems, the traditional seed supply systems are likely to be disrupted even more. It is, therefore, essential to study, document and embark on enhancing such systems, building on the above mentioned areas of community seed storage, use and exchange activities to develop sustainable sources of seed operating in networks, in a more coordinated and organized way.

WHAT IS A COMMUNITY SEED BANK?

Community seed banks are often understood as community-based stores used for the distribution of seed and grain to the local communities on a loan basis. In some cases, they are designed as income generating operations where high external input seeds with chemical packages are distributed to the farming community. But, as already discussed above, a community seed bank system is and should be a part of a community-managed genetic resources conservation and utilization practice (Regassa Feyissa, 2000). It is an integral part of an overall community-driven crop production strategy which farming communities have developed as part of their traditional farming systems. Within these systems, community-managed seed banks and on-farm farmers' variety maintenance are important components that serve as a source of sustained seed supply, as well as genetic materials for improved cultivar selection and enhancement.

The community seed bank described below represents a strategy for, or a collective approach to, the maintenance of genetic diversity in crop/plant species which also serves as a back-up for local self-sufficiency in planting material by stabilizing the seed supply system in cases of crop failure. It is a repository of locally adapted crop diversity, including enhanced farmers' varieties that are competitive in yield and other desirable characteristics with high input varieties that can be poorly adapted to local conditions.

Low-cost community level seed storage facilities can help to preserve the drought and climate change mitigating characteristics of traditional varieties, while, at the same time, serving as base material for farmers to select special lines to meet their changing needs. They also play a key role in improving market outlets through enabling communities to produce crops of known quality and in stabilizing prices over changing situations. Thus, community seed bank development contributes toward promoting economic empowerment of farmers.

THE COMMUNITY SEED BANK COMPLEX

Figure 1 illustrates a scheme for a comprehensive seed supply system that could be introduced across Sub-Saharan Africa. It has been developed from a schematic plan originally proposed for Ethiopia (Melaku Worede, 1997; Melaku Worede *et al.*, 1999).

The seed bank is at the centre of the seed network and offers various community services such as seed security (storage), seed distribution and exchange, germplasm restoration and introduction. It is the key component of the community seed network, representing a low-cost and low-technology demanding system that may be owned and managed by local communities as part of existing community services including cooperatives.

The seed bank proper comprises several major components: a seed store, a germplasm repository (for local crop improvement), a herbarium and a documentation section for holding records and information on local and scientific knowledge, and an administrative and records unit.

The seed store represents a seed reserve system consisting largely of local varieties, including those enhanced and/or selected and multiplied on-farm through either participatory plant breeding (PPB) and/or participatory variety selection (PVS), as well as locally adapted and adopted introductions obtained by way of exchange or from various other sources (gene banks, regional centres, donations, etc.).

Local farmers can have access to such materials on a loan basis, or through other arrangements as deemed appropriate for and by their community. The seed reserve also provides a back-stop to the local (informal) market networks where farmers traditionally exchange seeds and information.

Other components of such a seed bank system include botanical gardens, seed/plant micro-increase plots, and infrastructures for seed cleaning/processing, meeting/training facilities and, where feasible, a permanent source of water such as a well for multiple purposes. The catchment of water or other forms of water harvesting may also provide options for the supplementary watering needed for the botanical garden, off-season plant micro-increase plots, or even to ensure that plots of endangered varieties in very small quantities can mature if the rains are inadequate.

FIGURE 1

A diagram of a community seed bank complex and network for Africa

BREEDERS  GENEBANK

COMMUNITY

COMMUNITY

Botanical garden

Seed (bulk) cleaning, drying & storage

Meetings & training activities, also cultural events

Documentation (local knowledge & records) Administration

Supporting activities: soil & water conservation; cultural practices; cropping systems; marketing & financial services

Seed multiplication, distribution & exchange

ADVOCACY

Seed reserve

Germplasm repository

ON–FARM CONSERVATION, LOCAL STORAGE UNITS

PPB/PVS

Small tools/ other equipment

Water well or harvesting

Micro increase of seed

COLABORATIVE ACTIVITIES/SERVICES

SEED BANK

Networking (local national & regional)

LOCAL MARKETS

PPB = participatory plant breeding
PVS = participatory variety selection

Community activities for germplasm conservation and seed processing at the seed bank (germplasm repository) may also include surveying/collecting and characterization of seed, involving gene bank experts and modern and traditional plant breeders. Training of local community members should be provided to carry out routine activities.

Traditional storage units such as the small storage bins, clay pots, rock-hewn mortars and undergrounds pits maintained by local farmers, which form an integral part of the local traditional seed storage systems, may be linked to the community seed bank. Improved versions of these small units, as well as materials maintained as living collections *in situ* in the field and in backyard gardens should also be considered for such purposes.

The on-farm conservation and improvement of farmers' varieties included in the community seed bank complex require a good understanding of the distinct challenges in potential programme areas, prior to initiating any activities. Farmers' objectives and needs for seed vary according to the local farming system, and this determines the strategy for on-farm management of crop diversity (Regassa Feyissa, 2000).

In cases where traditional crop genetic diversity still exists on farms, a major step in undertaking the crop management activities is to develop competitively productive forms of farmers' varieties primarily through farmer-led participatory varietal selection and/or participatory plant breeding in order to improve production and ensure conservation through continued use. It is understood that such production improvements may entail some risks of genetic erosion unless careful and systematic conservation measures are taken.

In the case where genetic diversity is threatened and farmers' options are shrinking, as has been happening in many parts of Africa for example, the main objective is to restore the lost diversity through reintroduction or exchange of material among communities and/or through national and regional gene banks, such as the SADC (Southern Africa Development Cooperation) Regional Genebank for Southern Africa. Such exchanges could be facilitated through the network of community seed bank complexes.

Through this approach, farmers will be able to continue to control their choice of crop types and cultivars while maintaining a conservation programme, and they will have ready access to planting material adapted to local growing conditions. Farmers will also be in a position to critically evaluate the relative merits of a wide range of cultivars, thereby limiting undue expansion of exotic cultivars that are costly and poorly adapted.

In order for the smallholder farmers to effectively exploit their resources and fulfil their productive potential through the seed network, they require a range of enabling services and conditions; and each service has to be sustainable. This may include measures for supporting community initiatives for sustainable financial services, establishing savings and credit cooperative societies and other forms of assistance which could be incorporated into existing community activities to sustain the seed banking, seed multiplication and distribution, processing and marketing of products.

Developing marketing strategies for farm produce and value-added products are essential for creating the incentives (or removing the disincentives) to grow a wide range of crops, and in enlarging the food resource base for both rural/farm and urban communities of target areas in order to improve the incomes of the providers of these resources.

It is equally crucial to encourage activities that ensure the sovereign rights of farmers to the materials developed and produced on their farms, or to the benefits that might accrue as a result of improved market potential (national/international) – thereby generating the financial resources needed to sustain crop productivity and improved livelihoods. This has been set out in the International Treaty on Plant Genetic Resources for Food and Agriculture (FAO, 2009). It is also important to facilitate farmer-to-farmer exchange of material and information within and among communities through networks. Farmers' systems of informal seed and information exchange provide a continuous flow of material and knowledge which go beyond geographical boundaries (often bypassing governments and formal arrangements). Such dynamic farmers' seed exchange systems are often linked to strong local plant genetic resource management and high rates of variety improvements.

The role of extension agents and grassroots workers (agricultural, socioeconomic, etc.) is also crucial in linking the development of community seed banks with other community initiatives including soil and water conservation, and understanding cultural practices in a co-operative manner.

Advocacy is another activity that is included in the seed supply network. It may be undertaken, indirectly or in a supportive manner, through community empowerment to enable farmers to improve productivity and market their products to generate sustainable incomes, have access to and control over their sources of seed and means of production, control over the use of related plant genetic diversity, etc. Concurrently, the information (facts and figures) and knowledge generated in the process provides an invaluable backup for advocacy at all levels.

NETWORKING COMMUNITY SEED BANK ACTIVITIES AT NATIONAL, REGIONAL AND GLOBAL LEVELS

There is a growing worldwide concern and interest to promote ecological agriculture in the developing world as this is seen as being able to assist poor smallholder farmers achieve sustainable development beyond their current subsistence level. Central to this move is the recognition of the key role played by indigenous seeds and traditional farming practices. It is therefore, essential to network with existing initiatives in these areas both within Africa and other developing regions, sharing experiences on the conservation and effective utilization of the rich inheritance of crop genetic resources still found with these smallholder farmers.

Developing a community seed bank complex at various strategically selected locations within the African region where community seed supply projects already exist offers a valuable starting point for achieving this task. This will provide an opportunity for networking and coordination of community seed bank activities in target areas, as well as the platform for case studies and learning or awareness creating activities at local, national and regional levels.

There is one such project already being developed in South West Wello, Ethiopia within the community seed supply programme of EOSA, supported by Seeds of Survival/International (SOS/I) and Umanitarian Service Committe of Canada (USC) (Anonymous, 2009).

Finally, success in implementing such a network of community seed banks will depend largely on the willingness of agricultural professionals and policy-makers to learn from farmers, the living repositories of indigenous knowledge, and in no small measure on close partnerships and collaboration between scientists and farmers to achieve a synthesis between modern and indigenous knowledge, thereby creating a new knowledge base for sustainable development (Melaku Worede *et al.*, 2000). The synergy resulting from the combined use of scientific and farmers' know-how is key to the enhanced management of natural resources that enables farmers to produce food crops beyond the subsistence level for national food security as well as to meet the challenges of climate change.

REFERENCES

Anonymous. 2009. *Community-based diversity conservation (CBDC): Africa experiences, Phases III Report: 2007–2009.*

Emmanuel O. A., Asisey Mangetane, G. Khelikane & Melaku Worede. 1999. Links between indigenous knowledge and modern technology: Seeds of Hope, African perspectives – practices and policies supporting sustainable development. *In* David Turnham (ed), *Scandinavian Seminar College*, 4, pp. 51–68.

FAO. 2009. International Treaty on Plant Genetic Resources for Food and Agriculture. FAO, Rome.

Melaku Worede. 1988. Diversity and the genetic resource base [of Ethiopia]. *Ethiopian Journal of Agric. Sci.* Vol. 10, No. 1–2, pp. 13–14.

Melaku Worede. 1991. Crop genetic resources conservation and utilization: An Ethiopian perspective. *Science in Africa, Achievements & Prospects.* Proceedings of the Symposium of the American Association for the Advancement of Science, 15 February 1991, Washington D.C.

Melaku Worede. 1993. The role of Ethiopian farmers in the conservation and utilization of crop genetic resources. *In* D.R Buxton, R. Shibles, R.A.Forsberg, B.L. Blad, K.H. Asay, G.M.Paulsen & R.E.Wilson (eds), Annual Meeting of the Int. Crop Sci. Society of America, Madison,WI. pp. 290–301.

Melaku Worede. 1997. Ethiopian *in situ* conservation. *In* N. Maxted, B. V. Ford-Lloyd & J.G. Hawkes (eds). *Plant Genetic Conservation*: *The* in situ *approach*. pp. 290–301. Chapman and Hall, London.

Melaku Worede, Tesfaye Tessema & Regassa Feyissa. 2000. Keeping diversity alive: An Ethiopian perspective. *In* S.B. Brush (ed), *Genes in the field, on farm conservation of crop diversity.* pp. 143–161. IPGRI (International Plant Genetic Resources Institute), Rome, IDRC (International Development Research Centre), Ottawa, Lewis Publishers, London, New York, Washington D.C.

Regassa Feyissa. 2000. Community seed banks and seed exchange in Ethiopia: A farmer led approach. *In* E. Friis-Hansen and B. Sthapit (eds), *Participatory approaches to the conservation and use of plant genetic resources.* pp. 142–148. IPGRI, Rome.

HOW TO MAKE AND USE COMPOST

Sue Edwards and Hailu Araya

CONTENTS

TABLES

BOXES

FIGURES

ABOUT THE AUTHORS

SUE EDWARDS is the Director of the Institute for Sustainable Development (ISD), Addis Ababa, Ethiopia, and has been co-editor of the eight-volume "Flora of Ethiopia and Eritrea" since 1984. She is a taxonomic botanist, teacher and science editor by profession.

HAILU ARAYA is the Sustainable Community Development team leader of the ISD. He is a geographer who joined ISD after completing his Masters degree in community resource management at Addis Ababa University in 2001, and, in 2010, obtained a doctorate in soil science from University of Hohenheim, Germany.

ACKNOWLEDGEMENTS

The Institute for Sustainable Development (ISD) wishes to thank the Third World Network, based in Penang, Malaysia, for its consistent support since 1996 to the Project on "Sustainable Development through Ecological Land Management by Some Rural Communities in Tigray, Northern Ethiopia", also known as the "Tigray Project". The information in this document is based on experiences in training farmers and their associated agricultural professionals from the Bureau of Agriculture and Rural Development of Tigray in making and using compost for use with field crops. ISD also promotes and trains students in environment clubs in government schools as well as self-organized youth groups in organic vegetable gardening and the raising of tree seedlings. The feedback from all these partners of ISD has helped improve the present version. The original drawings were made by Solomon Demlie and edited with PhotoShop by Martha Mesfin and Samuel Tesfaye.

EDITORIAL NOTES

The information in this guide to making and using compost has been developed from working with Ethiopian smallholder farmers since 1996, particularly in the dry and degraded highlands of northern Ethiopia. It is based on the Tigrinya booklet by Arefayne Asmelash (1994 EC/2002 GC), the ISD Project Officer based in Mekele, Tigray. It is hoped that smallholder farmers and local agricultural experts in many parts of the world, and particularly in Sub-Saharan Africa, will be able to identify and use the most appropriate and applicable method for making compost in their own areas.

NATURAL FERTILIZER AND HEALTHY SOIL
Fertilizers

Natural fertilizer provides the food needed for a plant to grow after a seed has germinated in the soil. This food consists of plant nutrients. The most important of these nutrients are nitrogen (N), phosphorus (P) and potassium (K). There are also many other chemicals needed by plants in small quantities, e.g. copper (Cu), manganese (Mn), magnesium (Mg), iron (Fe), sulphur (S) and others. These are called micronutrients or trace elements. Natural fertilizer also provides organic matter called humus for the soil. Humus is a black or brown spongy or jelly-like substance. It helps the soil have a good structure to hold water and air. One of the best natural fertilizers is mature compost because it feeds the soil with humus and plant nutrients. The growing plants take their nutrients from the top layer of the soil where their roots grow.

Plant nutrients are lost from the soil when they are washed down (leached) below the top soil, or when the top soil is eroded. Plant nutrients are also lost with the crops when these are harvested. When the surface of the land is broken up for farming, the soil is often eroded: it is blown away by the wind or washed away by rain and floods. The soil also loses much of its carbon content as carbon dioxide (CO_2) into the atmosphere, thus contributing to climate change. The soil that is left becomes poor in plant nutrients so the crops do not grow well and give a good yield. But if the plant nutrients and carbon are returned to the soil, it can continue to grow good crops as well as contribute to slowing down the negative impacts of climate change.

Farmers can replace the lost plant nutrients by using fertilizers. Natural fertilizer comes from animal wastes and plants; for example, cow dung, sheep, goat or chicken droppings, urine, decomposed weeds and other plant or animal remains, e.g. waste from preparing food. The fertilizer can also be made of chemicals in a factory. Farmers have to buy this type of fertilizer from the market or through farmers' service cooperatives.

Therefore fertilizers are of two types:

- Natural fertilizer, including compost; and
- Human-made chemical fertilizer.

Throughout the world there are many options for replacing the plant nutrients lost from soil, but, in our case in Ethiopia and in many other parts of Sub-Saharan Africa where most of the agriculture is done by smallholder farmers, the best option is compost produced by human labour using the natural materials available to farmers and others, such as students and youth, from their surroundings. Good quality compost can be made from organic household wastes in urban areas and be used to grow healthy vegetables in gardens at home or by school environment club or youth group members.

Soil

The soil is a complex mixture of the following:

- Non-living materials – solid particles from broken down rocks, air and water;
- Living organisms – bacteria, fungi, many small and very small (microscopic) animals, plants such as algae and plant roots; and
- The decayed and decomposed remains of living organisms – humus.

The solid particles provide the basic structure or skeleton of the soil. Generally three types of particles are recognized: sand, silt and clay. Sandy soil is rough to feel because it is made of large grains. Sandy soil does not hold much water. Silty soil is finer to feel than sandy soil. When it is moist, the particles stick together in crumbs. Clay soil is very soft when wet as the particles are very small. They stick together even when the soil is dry and hard. Clay particles swell when they get wet and water cannot pass through easily (see Figure 1).

Natural soils consist of combinations of sand, silt and clay. The sand holds some plant nutrients and helps provide good drainage of excess water from a soil. Silt holds more plant nutrients and helps to hold water in the soil. Clay holds even more plant nutrients and water, but has little air. Loam or loamy soil contains a

FIGURE 1

Types of soils showing different texture

sandy soil   silty soil   clay soil

balance of sand, silt and clay. In a healthy soil, all these particles are coated with a layer of humus. This gives the soil its brown colour, good smell and structure. The humus also holds and helps keep plant nutrients and water in the soil.

Humus helps the soil particles stick together, but they do not fit tightly together. A loam soil with good humus has spaces or pores between the particles for water and air to get into and move through the soil. Humus is important for a soil because it:

- Holds moisture;
- Holds nutrients;
- Allows air to get into the soil; and
- Contributes to a good soil structure.

A healthy soil contains about 12–20 percent carbon, i.e. organic matter. The organic matter is the source of energy for the bacteria, fungi and other organisms in the soil. These organisms break down dead plant and animal remains releasing carbon dioxide, water and mineral salts, including nitrates, phosphates, etc., which are the nutrients for growing plants.

Some of the water in the soil is held tightly by the soil particles, especially by the clay, and plants cannot use it. Other water moves more freely through the pores, and this is available for plant growth. Humus acts as a water reservoir for the plant roots and other organisms in the soil. It can hold up to six times its own weight in water.

FIGURE 2

Some examples of soil organisms that help make compost and keep soil healthy

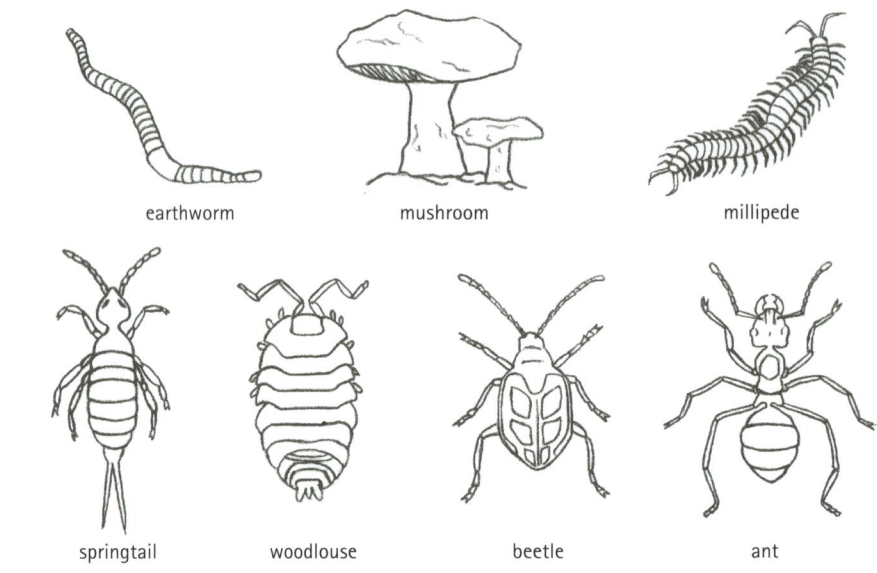

earthworm   mushroom   millipede

springtail   woodlouse   beetle   ant

The air in the soil has much more carbon dioxide than the above ground atmosphere. This is because the plant roots and the other living things in the soil produce carbon dioxide when they 'breathe', but the movement of air in the soil is slow and the carbon dioxide does not move out into the air as fast as from animals living above ground.

There are many organisms that live in the soil (see Figure 2). The bacteria and fungi are particularly important in breaking down plant and animal waste materials, and making plant nutrients available. Many fungi and bacteria also help in transferring nutrients from the soil to the roots of plants. The larger animals, worms, beetles, etc. help break down dead things into a condition that the bacteria and fungi can digest. These animals also move and mix the soil, sometimes dramatically like earthworms and termites. In a healthy soil, there is a very large mixed population of all these organisms. They each have a role to play in keeping the soil healthy, and hence, also the crops that grow on the soil. Pests are not usually a problem in a healthy soil. Thus, healthy soil produces healthy food.

THE CHARACTER OF COMPOST
Why is compost important?

Compost is important because it:

o Contains the main plant nutrients – nitrogen (N), phosphorus (P) and potassium (K), often written as NPK;

o Improves the organic matter in the soil by providing humus;

o Helps the soil hold both water and air for plants; and

o Makes trace elements or micronutrients available to plants.

What can compost be used for?

Because compost is made up of humus, it can be used for improving soil as follows:

1. It provides plant nutrients that are released throughout the growing season.

    o The plant nutrients are released when organic matter decomposes and is changed into humus.

    o The plant nutrients dissolve in the water in the soil and are taken in by the roots of the crops.

2. It improves soil structure so that plant roots can easily reach down into the soil.

    o In sandy soil the humus makes the sand particles stick together. This reduces the size of the spaces (pores) so that water stays longer in the soil.

    o In clay soils, the humus surrounds the clay particles making more spaces (pores) in the soil so the root systems of plants can reach the water and nutrients that they need, and air can also move through the soil.

    o Therefore, because heavy clay soils become lighter and sandy soils become heavier, soil that has had compost added to it is easier to work, i.e. to plough and dig.

3. It improves the moisture-holding capacity of soil.

    o The humus is a dark brown or black soft spongy or jelly-like substance that holds water and plant nutrients. One kilogram of humus can hold up to six litres of water.

- In dry times, soil with good humus in it can hold water longer than soil with little humus. In Ethiopia, crops grown on soil with compost can go on growing for two weeks longer after the rains have stopped than crops grown on soil given chemical fertilizer.
- When it rains, water easily gets into the soil instead of running off over the surface.
- Water gets into the subsoil and down to the water table, runoff and thus flooding is reduced, and springs do not dry up in the dry season.

4. It helps to control weeds, pests and diseases.
   - When weeds are used to make compost, the high temperature of the compost-making process kills many, but not all, of the weed seeds. Even the noxious weed, *Parthenium*, has most of its seeds killed when it is made into compost following the instructions given in this document.
   - Fertile soil produces strong plants able to resist pests and diseases.
   - When crop residues are used to make compost, many pests and diseases cannot survive to infect the next season's crops.

5. It helps the soil resist erosion by wind and water. This is because:
   - Water can enter the soil better and this can stop showers building up into a flood. This also reduces splash and sheet erosion.
   - Soil held together with humus cannot be blown away so easily by wind.

6. Compost helps farmers improve the productivity of their land and their income. It is made without having to pay cash or borrow money, i.e. farmers do not have to take credit and get into debt like they do for taking chemical fertilizer. But, to make and use compost properly farmers, either individually or working in groups, have to work hard.

FIGURE 3

Diagram showing the components of compost

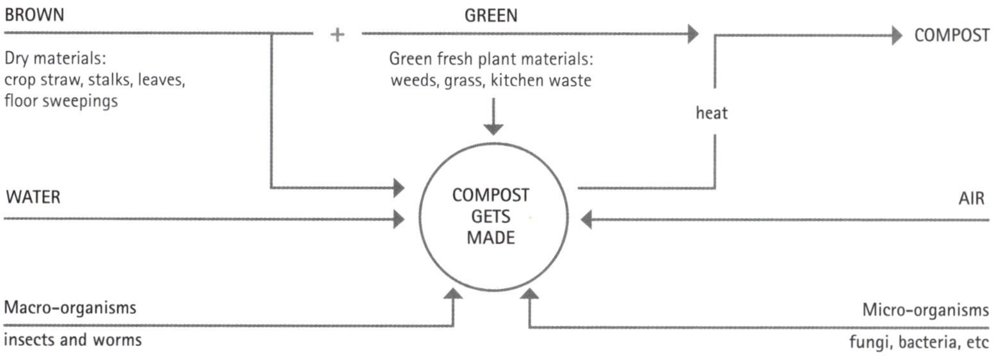

What is needed to make compost?
Plant materials, both dry and green

1. Weeds, grasses and any other plant materials cut from inside and around fields, in clearing paths, in weeding, etc.
2. Wastes from cleaning grain, cooking and cleaning the house and compound, making food and different drinks, particularly coffee, tea, home-made beer, etc.
3. Crop residues: stems, leaves, straw and chaff[1] of all field crops – both big and small – cereals, pulses, oil crops, horticultural crops and spices, from threshing grounds and from fields after harvesting.
4. Garden wastes – old leaves, dead flowers, hedge trimmings, grass cuttings, etc.
5. Dry grass, hay and straw left over from feeding and bedding animals. Animal bedding is very useful because it has been mixed with the urine and droppings of the animals.

---

1   Chaff = covering of grain crops left after threshing or pounding.

6. Dropped leaves and stems from almost any tree and bush except plants which have tough leaves, or leaves and stems with a strong smell or liquid when crushed, like *Eucalyptus*, Australian *Acacia*, *Euphorbia*, etc. However, we have found farmers making good quality compost including stems of *Euphorbia*.

7. Stems of cactus, such as prickly pear, can be used if they are crushed or chopped up. They are also a good source of moisture for making compost in dry areas. When the compost is made correctly, the spines are destroyed.

Water

Enough water is needed to wet all the materials and keep them moist, but the materials should not be made too wet so that they lack air and thus rot and smell bad. Both too little and too much water prevent good compost being made.

Water does NOT need to be clean like drinking water.

It can come from:

- Collected rainwater;
- Collected wastewater, e.g. from washing pots and pans, clothes, floors, etc.;
- Animal urine; or
- Human urine.

Water can also be collected from ponds, dams, streams and rivers, particularly if men are willing to do it. It is not fair to expect women to collect all the water needed to make compost.

Animal materials

1. Dung and droppings from all types of domestic animals, including from horses, mules, donkeys and chicken, from night pens and shelters, or collected from fields.

2. Chicken droppings are important to include because they are rich in nitrogen.

3. Urine from cattle and people:

- Catch urine in a container from animals when they wake up and start moving around in the morning.
- Provide a container – like an old clay pot or plastic jerry can – in the toilet or latrine where people can pass or put their urine.
- Night soil (human faeces): almost all human parasites and other disease organisms in human faeces are killed by the high temperatures when good compost is made.

Compost making aids – "farmers' friends"

Micro-organisms (fungi and bacteria) and smaller animals (many types of worms, including earthworms, nematodes, beetles and other insects) turn waste materials into mature compost. These are found naturally in good fertile soils like those from forests, old animal dung and old compost. Adding any of these to new compost helps in the decomposition process.

Adding compost making aids is like adding yeast to the dough to make bread. The farmers in Ethiopia call these materials the 'spices' to make good compost.

Air

Including dry materials in the compost, e.g. old leaves and stalks, provides space for air to circulate inside the compost. Air is needed because the soil organisms need oxygen.

Heat

Decomposition of organic wastes produces heat. Compost needs to be kept hot and moist so the plant and animal materials can be broken down quickly and thoroughly. Heat destroys most of the weed seeds, fungal diseases, pests and parasites.

The contributions of the different compost-making materials
A good balance of carbon and nitrogen

Both carbon and nitrogen are needed to make good compost. They are used by the micro-organisms to grow and multiply, and to get energy. Some of the carbon is converted to carbon dioxide, and this escapes to the atmosphere. Most of it remains and becomes humus, and the nitrogen becomes nitrates. Methane is not produced if there is a good supply of air to the organisms carrying out the decomposition process.

Materials with good nitrogen content help in making good compost, but they should be less than the carbon-containing materials. Carbon-containing materials should always be more than those containing high nitrogen. A good balance of carbon and nitrogen is needed to make good compost. Table 1 gives the carbon-to-nitrogen balance for some types of composting materials.

TABLE 1

The nitrogen and carbon content of some selected composting materials

TYPE OF COMPOSTING MATERIAL	NITROGEN CONTENT (%)	CARBON-TO-NITROGEN RATIO (C:1N)
Urine	15–18	0.8:1
Blood	10–14	3:1
Horn	12	not found
Bone	3	8:1
Chicken manure	3–6	10–12:1
Sheep manure	3.8	not found
Horse and donkey manure	3.8	25:1
Manure in general	1.7	18:1
Manure from animal pens = farmyard manure (FYM)	2.15	14:1
Maize stalks and leaves	0.7–0.8	55–70:1
Wheat straw and chaff	0.4–0.6	80–100:1
Fallen leaves	0.4	45:1
Young grass hay	4	12:1
Grass clippings	2.4	20:1
Straw from peas and beans	1.5	not found

Sources: Dalzell and Riddlestone (1987), Gershuny and Martin (1992)

With Nitrogen as 1, high figures for the carbon in the carbon-to-nitrogen column indicate high carbon content. These items are good for making compost. Items with low carbon content, like urine and chicken manure, are useful to provide nitrogen. But they must be mixed with materials with high carbon content.

1. When there is enough air and moisture in the compost, nitrogen-containing materials are broken down and the nitrogen is changed to nitrates that can be used by plants.

2. When there is too much water and little air, the nitrogen is changed into ammonia. This is a gas that escapes from the compost, and gives the compost a bad smell.

3. When there is a bad smell, the compost needs to be turned over bringing the top to the bottom and the bottom to the top, and mixing in more dry materials and some good soil. This puts more air into the compost, which stops the process of making ammonia so that proper mature compost can be made.

The contributions of dry and green plant materials

Dry materials give structure to the compost making process; they provide space for air to circulate so that the micro-organisms can be active and make heat.

Green plant materials provide moisture for compost making; they give water and nutrients to the micro-organisms so that they multiply and break down the organic materials into humus.

The importance of good water/moisture and air balance

Water is essential for compost preparation.

1. Sufficient moisture helps for quicker decomposition because it is essential for micro-organisms to be active.

2. Excess water causes rotting of the materials and creates a bad smell.

3. Without enough moisture the decomposition process slows down and the materials will not be changed into compost.

This shows that moisture and air must be balanced to make good compost. Farmers quickly learn how to judge the amount of water needed to be added in making compost.

The importance of air

Compost should have sufficient air.

1. When there is sufficient air, oxygen enters the compost heap. When there is enough oxygen, special bacteria can convert nitrogen into nitrate, the materials are decomposed properly and there is a good smell.
2. If there is not enough air and too much water, the nitrogen is converted into ammonia. The ammonia escapes into the air removing nitrogen from the compost and making it smell bad.
3. If there is excess air and too little water, the materials dry up and do not decompose to become compost.

BOX 1
EXAMPLES OF SOME PLANT MATERIALS FOR MAKING COMPOST

Crop straws absorb water without changing their physical structure. They are good for keeping air in the compost, but they do not mix easily with other materials and decompose slowly.

Grass and other green materials have usually lost water and wilted before they are put into the compost. They can hold moisture longer in a compost pit than in a compost heap.

Farmers in Ethiopia have found it is best to thoroughly mix the dry and green plant materials together before they are put into the compost pit or compost heap.

Quality compost with animal dung and urine

1. Animal dung contains water, nitrogen, phosphorous and potassium, as well as micro-nutrients.
2. Animal dung and urine are very necessary to prepare good quality compost – urine especially is high in potassium and nitrogen.
3. Both dung and urine help to produce a high temperature so that the materials decompose into compost easily.
4. Urine, in particular, accelerates decomposition.

Important compost making aids

Compost making aids are farmers' friends as they help speed up the process of decomposition. They are like the yeast in making bread and beer or wine, or the salt and spices in making tasty food. They include:

1. **Good top soil and old compost**: These contain bacteria, fungi and many small animals to work on breaking down the materials into mature compost.
2. **Ashes** from wood and charcoal are good to mix in because they contain phosphorous, potassium, and many micro-nutrients like zinc, iron and magnesium.
3. **Heat** is produced by the action of bacteria and fungi on the plant and animal materials, and their activity keeps the compost hot. Covering compost with a black plastic sheet can also absorb the heat from the sun and stop it escaping so that the compost making process goes fast.
4. **Larger organisms**: Look for larger organisms, like earthworms and beetles, in old moist compost, old animal dung or good top soil and add these to the compost making materials without drying and sieving them.
5. **Composting facilitators/promoters** are important because:
   - They provide key bacteria, fungi and micro-organisms to make the compost;
   - They provide nutrients for the organisms in the soil so they remain in a good condition and reproduce rapidly; and

○ They help speed up the composting process and ensure that good quality compost is produced.

Methods for using compost making aids include any or all of the following:

1. Make a mixture of dry top soil, old compost and ashes. Then crush it and, if possible, sieve it so it is like salt or a fine powder.
2. Mix the powder with fresh composting materials, particularly with dry or green plant materials like grass and/or straw, and put this in layers between other materials.
3. Do NOT put the compost making aid material as a layer by itself. It needs to be mixed with the other materials so it can accelerate the compost making process.
4. Ash is good as it contains minerals, BUT if you put a high quantity in one layer, the minerals are strongly concentrated and can slow down or stop the micro-organisms from making compost.

How micro- and macro-organisms work

The production of good quality mature compost depends on the number and types of micro- and macro-organisms living in the soil. These are living organisms that require air, moisture and heat in the compost heap so that they can live, work and multiply/reproduce.

Compost materials supply food and energy (starch, soluble sugars, carbohydrates, amino acids) for the micro-organisms.

In the presence of air supplying oxygen and moisture, the micro-organisms convert the available food into humus and soluble plant nutrients, which stay in the compost heap, and carbon dioxide, which diffuses out into the atmosphere. Most of the carbon in compost materials stays in the humus and only a small amount leaves as carbon dioxide.

As the micro-organisms grow and multiply, they produce heat which speeds up the compost making process. Heat also kills many weed seeds, pests, parasites and diseases from the fields, and in the animal dung and human faeces.

The heat ensures that healthy mature compost is produced.

CONDITIONS TO BE FULFILLED BEFORE PREPARING COMPOST

There are two main methods for preparing compost. One is called the Indore method and the other is the Bangalore method. The names come from districts in India where the compost making processes were first developed. The difference between the two methods is in the way the materials are put together and in the time taken for completing the compost heap or filling the pit.

The **Indore method** can be prepared either in a pit or as a heap or pile above the ground, but its preparation must be completed in less than a week.

The complete Indore method uses a sequence of three layers of materials: dry plant materials, green plant materials, animal manure and some soil. It is suitable for times and places where there are plenty of materials to make the mature compost, and labour, such as in a school or with a farmers' group, to put them together quickly.

The **NADEP method** is like the Indore method except that the tank is filled in one or two days and it always includes animal manure. This method needs a lot of work, but it produces very high quality mature compost without any more labour after the NADEP tank has been filled and sealed.

The **Bangalore method** is prepared in areas where composting materials and water availability are limited, and labour is also limited. The materials can be collected over a week or more, and then the new layers are made until either the heap is about 1 to 1.5 metre tall, or the pit is full. The Bangalore method uses only two layers of materials: dry plant materials and green plant materials. It is very suitable for making compost from household wastes, or in farms where there are no domestic animals.

Both the Indore and the Bangalore methods can include animal manure as an additional layer. Including animal manure ensures the best quality compost. But good quality compost can be made even without animal manure, i.e. just from plant materials and kitchen wastes.

Preparing compost needs dedication. Therefore:

○ Decide when and what method to use to make the compost.

○ Look out and search for composting materials that can be collected and carried to the compost-making place.

- Find out who will provide the water, and how.
- Decide if it is possible to collect and use urine.
- Be prepared to give time and effort, i.e. work hard, to prepare good quality compost.
- Set a target for the area of farmland or garden to be covered by the mature compost. Adding mature compost to a small field or even a small area in a field and then planting it with a high value crop can show good economic returns in a year.
- Collecting composting materials, layering or piling, and mixing are the main tasks during compost making. These need physical and mental preparation to overcome the burden of hard work, but it is only for a short time.
- Seeing good crops grow well and getting good yields from well-composted soil is very rewarding.

In Ethiopia, and other places with warm to hot climates, mature compost can be prepared in three to four months. In colder places, decomposition to make mature compost can take from six months to a year.

BOX 2
MAMMA YOHANNESU AND FINGER MILLET

Mamma Yohannesu was an old woman living with her grandson. She had a very small field of about 10 x 25 m. The soil was rather sandy. She managed to make about five sacks of compost, which she put on this field when her neighbour ploughed it for her. She planted the field with finger millet. In most of the field she scattered the seed, but in a plot of 5 x 5 m she brought and planted young finger millet seedlings she had grown in her house garden. She got a fantastic yield for her efforts – equivalent to 2.8 tonnes/ha for the directly sown finger millet and 7.8 tonnes/ha from the transplanted seedlings.

Points to remember when making compost in a heap

1. It is good to make a heap in the rainy season when there is plenty of green plants, such as weeds, getting water is easy or the materials are naturally wet, or where there is plenty of water available.
2. The compost heap will be on the ground with its base in a shallow trench to hold the foundation layer.
3. It should be in a place where it can be protected and get covered with leaves or straw or plastic during the rains so that the materials are not damaged or washed away.
4. It can be made under the shade of a tree and covered with wide leaves or plastic in order to protect the heap from high winds.
5. After the rains stop, keep the heap covered and check regularly to see if the moisture and temperature are correct, as described later in the section on follow-up.

Points to remember when making compost in a pit

1. This is good anytime of the year where moisture is limiting, and is the best way to make compost after the rains have finished and during the dry season.
2. Prepare and dig the pit, or better still, a series of three pits, when the land is moist and easier to dig, and/or when there is a gap between other farming activities.
3. If possible, make the compost immediately at the end of the rainy season while there are plenty of green and moist plant materials.
4. In the dry season, make the pit near a place where water can be added, e.g. next to the home compound where waste water and urine can be thrown on the compost materials, or near a water point, e.g. a pond, or near a stream where animals come to drink.
5. Mark the place of the pit with a ring of stones or a small fence so people and animals do not fall into it accidentally.

INDORE COMPOST PREPARATION METHODS

The Indore compost preparation method is done over a short period of time and in a systematic way of putting the materials together.

This method is most suitable for the rainy season when there are plenty of materials, e.g. weeds, to put into the compost. However, the place for making compost should be well-drained and easy to protect from floods and excess rain. The compost can be made either by piling in a heap or heaps, or in a pit or pits.

This method can also be used by vegetable growers when they clean their fields before the next crop is planted. The residues left after the crop is finished and harvested, such as stems and leaves from pumpkins, potatoes, tomatoes, chilli peppers and courgettes/zucchini, leaves and stalks from cabbage, etc. and any damaged crops that cannot be sold or eaten, can be collected together and organized for making compost.

Indore Piling Method
Selecting the site

The following factors need to be considered:
1. The site should be accessible for receiving the materials, including water and/or urine, and for frequent watching/monitoring and follow-up.
2. The site should be protected from strong sunlight and wind, e.g. in the shade of a tree, or on the west or north side of a building or wall.
3. The site should be protected from high rainfall and flooding.

Preparing the site

1. Clear the site of stones, weeds and grasses, but do not cut down any young trees. Instead, put the site so it is in the shade of the tree(s). The tree(s) will grow, provide shade and protect the compost heap.

2. Mark out the area for the compost heap. A minimum area is 1.25 m x 1.25 m. If it is smaller than this, the heap will dry out quickly so compost will not be made properly. The area can be larger, up to 3 m x 2.5 m.

3. Dig a shallow trench in the ground the same size as the compost heap. Make the trench about 20–25 cm deep. The bottom and sides of the trench should be smeared with water or a mixture of cow dung and water. This seals the pit so that moisture with nutrients does not leak out of the base of the compost heap.

4. The foundation layer of compost making materials is placed in the trench or pit.

5. The trench holds moisture during the dry season.

6. Materials are added in layers to make the heap, as shown in Figure 6 and described in more detail below.

The layers in making the compost heap

The foundation layer

1. Dry plant materials, e.g. strong straw and stalks of maize and sorghum, which are thick and long, are used for the foundation. These need to be broken into short lengths (about 10–15 cm). The stalks can be crushed, and then chopped. If possible let cattle lie down or sleep on them for one night. Walking cattle over the stems and stalks, as in threshing, is a good way of breaking up the stalks.

2. Spread the dry materials evenly over the bottom of the trench to make a layer 15–25 cm thick, as deep as a hand. Then sprinkle water with a watering can or scatter water evenly by hand over the dry plant materials so they are moist, but not wet.

3. The foundation layer provides ventilation for air to circulate, and excess water to drain out of the upper layers.

The three basic layers

1. The compost heap is built up of layers of materials, like in a big sandwich. The basic sequence is:

- **Layer 1:** A layer of dry plant materials, or mixture of dry plant materials with compost making aids like good soil, manure and/or some ashes. The layer should be 20–25 cm thick, i.e. as deep as a hand. The compost making aids can be mixed with the water to make slurry. Water or slurry should be scattered by hand or sprinkled with a watering can evenly over this layer making it moist but not soaking wet.

- **Layer 2:** A layer of moist (green) plant materials, either fresh or wilted, e.g. weeds or grass, plants from clearing a pathway, stems and leaves left over from harvesting vegetables, damaged fruits and vegetables. Leafy branches from woody plants can also be used as long as the materials are chopped up. The layer should be 20–25 cm thick. Water should NOT be sprinkled or scattered over this layer.

- **Layer 3:** A layer of animal manure collected from fresh or dried cow dung, horse, mule or donkey manure, sheep, goat or chicken droppings. The animal manure can be mixed with soil, old compost and some ashes to make a layer 5–10 cm thick. If there is only a small quantity of animal manure, it is best to mix it with water to make slurry, and then spread it over as a thin layer 1–2 cm thick.

2. Layers are added to the heap in the sequence, Layer 1, Layer 2, Layer 3, until the heap is about 1–1.5 metres tall. The layers should be thicker in the middle than at the sides so the heap becomes dome-shaped. If the heap is much taller than 1.5 metres, the microbes at the bottom of the heap will not be able to work well.

3. Layers 1 and 2 are essential to make good compost, but Layer 3 can be left out if there is a shortage or absence of animal manure.

4. Place one or more ventilation and/or testing sticks vertically in the compost heap remembering to have the stick long enough to stick out of the top of the heap. Ventilation and testing sticks are used to check if the decomposition process is going well, or not. A hollow stick of bamboo grass (*Arundo donax*) or bamboo makes a good ventilation stick as it allows carbon dioxide to diffuse out of and oxygen to diffuse into the heap. A testing stick is needed as it can be taken out at regular intervals to check on the progress of decomposition in the heap.

Making the covering layer

The finished heap needs to be protected from drying out, and also from animals pushing into it and disturbing it.

1. The covering layer can be made of wet mud mixed with grass or straw, with or without cow dung, or wide leaves of pumpkin, banana, fig trees, etc., or from plastic, or any combination of these materials, i.e. mud plaster covered with leaves or plastic, or leaves covered with plastic.

2. The cover should be put on both the sides and the top of the heap with only the ventilation stick coming out of the top.

3. The covering layer:
   - Prevents rain water from getting into the heap and damaging the compost making process; and
   - Helps keep heat inside the compost making heap. See the section on follow-up for how to check on the heat and moisture in the compost.

4. The compost heap can also be protected by putting a ring of stones or making a small fence around it.

5. The compost heap is best left untouched until there is mature compost inside it, or it can be turned over, as described for the pit method. If the compost is turned over, water should be sprinkled over the layers to keep all the materials moist. It is not necessary to try and keep the original different layers when turning over the compost – it is best if all the materials can be well mixed together, then added in layers about 20–25 cm thick and water sprinkled or splashed over them.

6. A mature compost heap is about half the height of the original heap, and the inside is full of dark brown or black substance, humus, which smells good. When the compost is mature, it should be very difficult to see the original materials.

7. This mature compost can be used immediately in the field, or it can be covered and stored until the growing season. When it is put in the field, it should be covered quickly by soil so the sun and wind do not damage it, and the nitrogen

does not escape to the atmosphere. Therefore, it is best to put compost on the field just before ploughing, or at the same time as sowing the crop. For row planted crops, it can be put in the furrow with the seed. For transplanted crops, it can be put in the hole with the seedling.

Indore Pit Method

The Indore pit method is best done at the end of the rainy season or during the dry season. It is important to make the pits where there is sufficient water available; for example, by a pond, small dam, run-off from a road or track, etc. Women should not be expected to carry water just for making compost. Waste water and urine from people and animals can be collected in old containers, and used in making compost.

The main reasons for making pit compost in the dry season are as follows:

1. After harvesting is complete, farmers can arrange their time to make compost including working together in groups according to their local traditions to share their labour.

2. If farmers have a biogas digester, the bioslurry from the digester can be used to make high quality compost at any time of the year, but particularly during the dry season.

3. The pits can be filled two or more times so that a large quantity of compost can be made over the duration of the dry season.

4. If pit compost is made during the rainy season or in very wet areas, water can get into the bottom of the pit. This will rot the materials producing a bad smell and poor quality compost. In wet areas it is better to make compost through the piling method.

5. Poor quality compost will not be productive and this can discourage farmers and others from trying to make better quality compost.

6. It is very important to have frequent follow-up and control of the balance of air and water in the materials being decomposed to make compost.

Selecting and preparing the site

1. The site should be accessible for receiving the composting materials, including water and urine, and for frequent watching/monitoring and follow-up.
2. The site should be protected from strong sunlight and wind. It should be in a protected area, for example, in the shade of a tree, or on the west or north side of a building or wall.
3. The pit or pits should be marked or have a ring of stones or small fence around it or them so that people and animals do not fall into it or them.
4. The site should NOT be where floods can come.

Digging the pits

The aim is to have a series of three pits, one next to the other. The minimum size for each pit should be:

- 1 metre deep (pits should NOT be deeper than 1 metre);
- 1–1.5 metres wide;
- 1–1.5 metres long (or longer).

However:

- The pits can be dug as they are needed – see Table 2 showing the flow of work.
- If a farmer and his/her family feel they have limited capacity, they can dig one pit of the above size, but then they should probably make compost using the Bangalore method (see next section).
- Smaller pits usually dry out too quickly so good quality compost is not made, and this will discourage the farmer from making and using compost.
- Pits deeper than 1 metre can be cold at the bottom and the micro-organisms cannot get enough oxygen to work properly.
- If compost is prepared by a group of farmers or students in an environment club or youth group, they can make a wider and/or longer pit that can supply all the families in the group. It also depends on the amount of composting

materials they are able to collect and bring to the pit. See also the sections on trench composting and the NADEP method, which are more suitable for compost making by groups, and where large quantities of composting materials are readily available.

- After the pit or pits are dug, they should be checked carefully to make sure there is no leakage of water into the pit which could spoil the compost making process.

Layers for filling the pit

Before the pit is filled, the bottom and sides should be covered with a mixture of animal dung and water – a slurry. If animal dung is not available, a mixture of top soil and water can be used. This plaster helps seal the sides of the pit so that moisture stays in the compost making materials.

The foundation layer

1. Dry plant materials, e.g. strong straw and stalks of maize and sorghum, which are thick and long, are used for the foundation. These need to be broken into short lengths (about 10–15 cm long). The stalks can be crushed, and then chopped. If possible let cattle lie down or sleep on them for one or two nights. Walking cattle over the stems and stalks, as in threshing, is a good way of breaking up the stalks. The cattle will add their dung and urine to the stalks making them more valuable for making compost.
2. Spread the dry materials evenly over the bottom of the pit to make a layer 20–25 cm thick. Then sprinkle water with a watering can or scatter water evenly by hand over the dry plant materials so they are moist, but not wet.
3. This is a very important layer in making pit compost as it makes sure that air can circulate through to the bottom of the pit.

The three basic layers

1. The compost pit is filled with layers of materials, like in a big sandwich. The basic sequence is:
   - ○ **Layer 1**: A layer of dry plant materials, or mixture of dry plant materials with compost making aids like good soil, manure and/or some ashes. The layer should be 20–25 cm thick, i.e. about the depth of a hand at the sides. The compost making aids can be mixed with the water to make slurry. Water or slurry should be scattered by hand or sprinkled with a watering can evenly over this layer. The layer should be moist but not soaked.
   - ○ **Layer 2**: A layer of moist (green) plant materials, either fresh or wilted, e.g. weeds or grass, plants from clearing a pathway, stems and leaves left over from harvesting vegetables, damaged fruits and vegetables. Leafy branches from woody plants can also be used as long as the materials are chopped up. The layer should be 20–25 cm thick at the sides. Water should NOT be sprinkled or scattered over this layer.
   - ○ **Layer 3**: A layer of animal manure collected from fresh or dried cow dung, horse, mule or donkey manure, sheep, goat or chicken droppings. The animal manure can be mixed with soil, old compost and some ashes to make a layer 5–10 cm thick. If there is only a small quantity of animal manure, it is best to make slurry by mixing the dung in water, and then spread it over as a thin layer 1–2 cm thick.

2. Layers are added to the pit in the sequence, Layer 1, Layer 2, Layer 3, until the pit is full to the top with the middle about 30–50 cm higher than the sides. The layers should be thicker in the middle than at the sides so the top becomes dome-shaped.

   Layers 1 and 2 are essential to make mature compost, but Layer 3 can be left out if there is a shortage or absence of animal manure.

3. Place one or more ventilation and/or testing sticks vertically in the compost pit remembering to have the stick long enough to stick out of the top of the pit. Ventilation and testing sticks are used to check if the decomposition process is going well, or not.

A hollow stick of bamboo grass (*Arundo donax*) or bamboo makes a good ventilation stick as it allows oxygen to diffuse into the pit. A solid stick is important as it can be taken out every few days to check on the progress of decomposition of the materials in the pit.

Covering the pit

After the pit is full of compost making materials, the top should be covered with wet mud mixed with grass and/or cow dung, and/or wide leaves such as those of banana, pumpkin or even from fig trees, and/or plastic so the moisture stays inside the pit, and rain does not get in to damage the decomposition process.
Note: Mark the place and/or cover the top with branches so animals and people do not tread on the cover and break it.

The progress in making compost should be checked regularly by taking out the ventilation or testing stick and checking it for heat, smell and moisture. The inside of the pit should be hot and moist with a good smell. The top of the pit will also sink down as the composting materials get decomposed.

Turning over and making compost throughout the dry season

1. In warm climates, about one month after the pit has been filled the compost can be turned over and checked.
2. In cold climates, the compost making materials take two or more months to start to decompose well. The rate of decomposition can be checked through the use of the testing stick.
3. A good farmer or gardener will soon learn how to judge the best time to turn over her or his compost.

Table 2 and Figure 4 show the sequence of activities for digging, filling and turning over compost in the three-pit system. This system spreads out the work so that a farmer who wants to have a good quantity of quality compost can plan and prepare it before the growing season.

TABLE 2

Sequence of activities for digging, filling and turning over compost in pits

PITS	PIT A	PIT B	PIT C	STORING OR USING MATURE COMPOST
1st month	Dig pit A Fill pit A with compost materials			
2nd month	Fill pit A for a second time	Dig pit B Put compost materials from pit A into pit B		
3rd month	Fill pit A for a third time	Put compost materials from pit A into pit B	Dig pit C Put compost materials from pit B into pit C	
4th month		Put compost materials from pit A into pit B	Put compost materials from pit B into pit C	Use mature compost from pit C or store it in pit C
5th month			Put compost materials from pit B into pit C	Use mature compost from pit C or store it
6th month				Use mature compost from pit C or store it

FIGURE 4

Diagram showing the sequence of filling pits A, B and C with compost materials

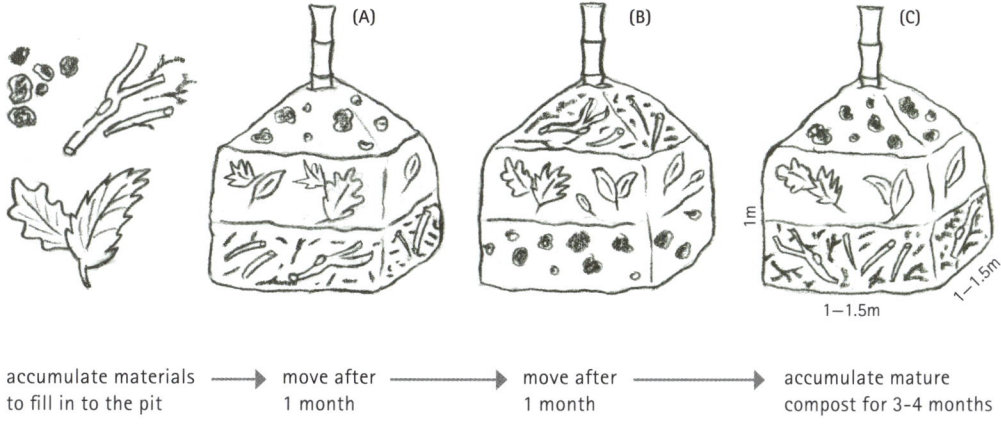

accumulate materials ⟶ move after ⟶ move after ⟶ accumulate mature
to fill in to the pit   1 month   1 month   compost for 3-4 months

The sequence for making a good quantity of quality compost using the three-pit method is as follows:

1. The cover is removed and all the materials are turned over into the second pit, i.e. from pit A to pit B. It is important to put the materials from the top of pit A into the bottom of pit B, and so on, with the materials from the bottom of pit A getting to the top of pit B. The materials can be mixed together, but they should be added in layers 20–25 cm thick and sprinkled with water to make sure they stay moist, but NOT soaked.

2. At the same time check that the moisture and air balance is correct. If the materials are too dry, more water should be sprinkled over them as they are put into the pit. If the materials are too wet, add more dry plant material in layers between the wet decomposing materials.

3. If the compost making is going well, you will find that the materials from pit A do not completely fill pit B. You will also see the white threads of fungi and many kinds of small organisms that are living on and decomposing the composting materials. The composting materials will have started to turn dark brown or black.

4. Pit A can now be filled for a second time with a new lot of composting materials as described above. Both pits should be closed with a layer of mud or leaves and/or plastic, as described above.

5. Again after about another month, the cover over pit B can be opened and the materials turned into pit C, and the cover to pit A removed and the materials in pit A turned over into pit B. At the same time check that the moisture and air balance in the materials is good.

6. If the compost-making process is going well, after two months the materials in pit B should be well decomposed, i.e. dark brown or black, with a good smell.

7. Pit A can now be filled for a third time with new composting materials, if they are available.

8. After a third or fourth month in warm climates, it should be possible to find fully matured compost in pit C. The material should look like good dark soil without any of the original materials visible. However, pit C may be only half full after the first lot from pit B is put into it. In fact, pit C can store all the compost until it is needed. Pit C should always be covered to prevent rain getting in, nutrients getting out and the compost being spoiled.

9. Or, the mature compost can be taken out, piled up and covered to be stored in a dry, cool and shady place until it is needed. It must be covered so that it does not blow away or the nutrients would be destroyed by sunlight or rain.

10. The mature compost can be taken out and put on the field just before ploughing, or mixed into the soil immediately by hand. The compost must be covered with soil so that the nutrients, particularly nitrates, are not destroyed by the sunlight.

With enough moisture and heat, compost making is fast under Ethiopian conditions. Four months after filling the first pit, it is possible to have compost to use on the land. By the sixth month, a good farmer can accumulate three lots of compost, enough for half a hectare of land.

BANGALORE COMPOST PREPARATION METHODS

The Bangalore method is not as precise or as demanding as the Indore method because the composting materials are added as they become available. It is highly suitable where there is a shortage of both composting materials and water. The Bangalore method can be used for both piling and pit methods, but the pit method is preferred in Ethiopia. This is because the pit holds moisture better than the heap, and the wind cannot blow away the materials so easily in the dry season. However, inside house compounds, the piling method is also convenient.

FIGURE 5

Sequence in making a compost heap

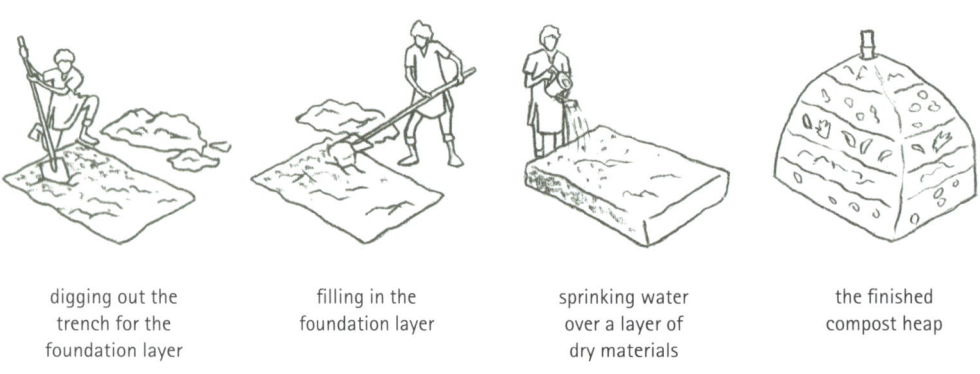

| digging out the trench for the foundation layer | filling in the foundation layer | sprinking water over a layer of dry materials | the finished compost heap |

Bangalore Piling Method (see Figure 5 above)
Selecting and preparing the site

1. Select a site where it is easy to add materials, e.g. inside a house compound.
2. The site should be sheltered from rain and wind. The best is in the shade of a tree, or on the north or west side of a building or wall to be sheltered from sun for most of the day.
3. Clear the site of stones and weeds, but leave trees to grow and give shade.
4. Mark out the length and width of the heap; for example, 1–2 m x 1–1.5 m and dig a trench 20–25 cm deep, i.e. about the depth of a hand, to be at the bottom of the heap to hold the foundation layer and stop it drying out in the dry season.

Making the heap

The foundation layer

1. Prepare the foundation layer from dry plant materials such as old straw, stalks of maize and sorghum, or old cabbage stalks, rose and hedge trimmings from gardens.
2. Use straw and maize and sorghum stalks as livestock bedding for one or two nights so that they get broken up and mixed with urine and dung.
3. Collect the materials and put them into the trench to make an even layer 15–25 cm deep. Sprinkle or scatter some water over the layer so it is moist but not wet.
4. Cover the layer with a little soil and some large leaves from banana, or pumpkin, or a fig tree, or even a sheet of plastic to prevent the materials drying out or being blown away.

Making the other layers

1. During the week, collect materials and put them in a convenient container such as an old jerry can, or next to the compost heap. Dry plant materials can be mixed with fresh moist ones, or the two types of plant material can be kept separately. The farmers in Ethiopia prefer to mix the dry and moist plant materials together. These materials can come from spoiled animal feed where animals have been stall fed, from cleaning the house and compound, clearing paths, weeding, stems and leaves after harvesting vegetables, preparing vegetables for making food, damaged fruits and vegetables, etc.
2. The dry materials can be used as livestock bedding for one or two nights so they collect urine and dung, and the animals can walk over them to break them up.
3. At the end of a week, remove the large leaves or plastic covering the top of the foundation layer so they can be used again, or leave the leaves to become part of the compost if they are too damaged to be used again.

4. Make a mixture of compost making aids like good soil, old manure and/or some ashes as a fine powder. Mix these with the dry plant material, or with the mixture of dry and moist plant material.

5. First add the layer of dry plant materials that have been used as bedding with the animal urine and dung in them, and then put the layer of green plant materials on top, OR add a layer of the mixed dry and moist plant materials. Make each layer 15–25 cm thick with the middle thicker than at the sides so that the heap becomes dome-shaped (see Figure 5).

6. Cover each layer with a thinner layer of animal manure or soil and/or big leaves like those from banana or pumpkin or fig trees so that the composting materials do not dry out. Animal manure can be left out if it is not easy to get, but the soil is important.

7. Repeat this process each week, or whenever there are enough materials collected to make one or two new layers, until the heap is about 1–1.5 metres tall. Make the centre of the heap higher than the sides so that the heap has a dome shape (see Figure 6).

8. Put a testing and/or ventilation stick into the middle of the heap.

Making the covering layer

The finished heap needs to be protected from drying out, and also from animals pushing into it and disturbing it.

1. The covering layer can be made of wet mud mixed with grass or straw, with or without cow dung, or wide leaves of pumpkin, banana, fig trees, etc., or from plastic, or any combination of these materials, i.e. mud plaster covered with leaves or plastic, or leaves covered with plastic.

2. The cover should be put on both the sides and the top of the heap with only the ventilation stick coming out of the top. The covering layer:
   o Prevents rain water from getting into the heap and damaging the compost making process; and
   o Helps keep heat inside the compost making heap. See the section on follow-up for how to check on the heat and moisture in the compost.

FIGURE 6

Diagram showing the layers in a Bangalore compost heap

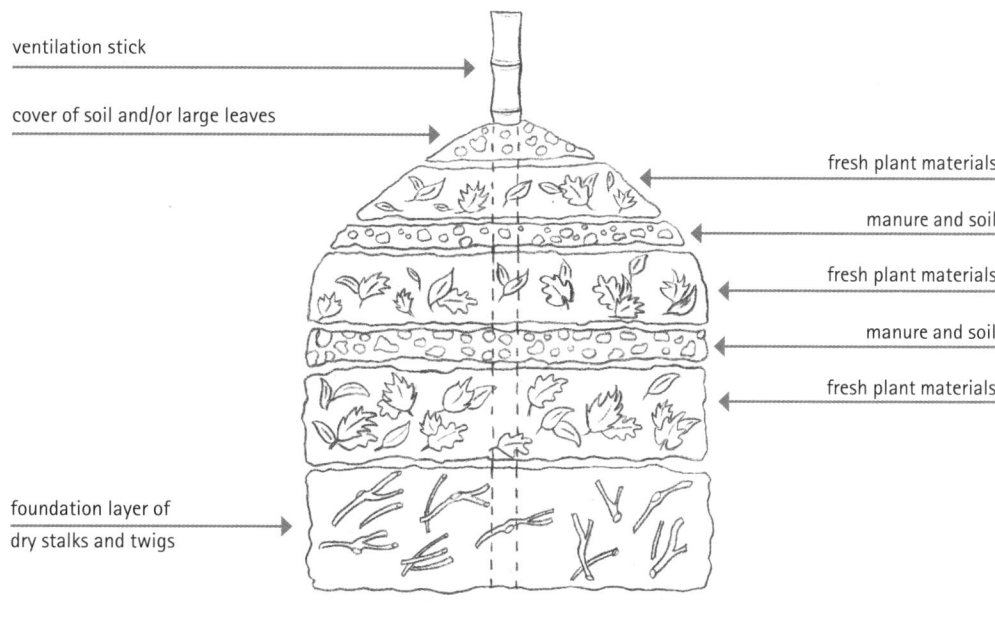

ventilation stick

cover of soil and/or large leaves

fresh plant materials

manure and soil

fresh plant materials

manure and soil

fresh plant materials

foundation layer of
dry stalks and twigs

3. The compost heap can also be protected by making a small fence around it from branches.

4. The compost heap is best left untouched until there is mature compost inside it, or it can be turned over, as described for the pit method. If the compost is turned over, water should be sprinkled over the layers to keep all the materials moist. It is not necessary to make the different layers when turning over the compost – all the materials can be well mixed together, then added in layers about 20–25 cm thick and water sprinkled or splashed over them.

Where the climate is warm, mature compost can be ready in about four months.

Bangalore Pit Method
Selecting and preparing the site

1. It should be in a place that is easy to take the materials, including water and urine, to the pit as well as for watching and follow-up.
2. The site should be protected from strong sunlight and wind. It can thus be, for example, in the shade of a tree, or on the west or north side of a building or wall.
3. The pit should be marked or have a ring of stones or a fence of branches around it so that people and animals do not fall into it.
4. The site should be protected and away from where floods can come.

Digging the pit

1. The minimum size of a pit should be:
   - 1 metre deep (pits should NOT be deeper than 1 metre);
   - 1–2 metres wide;
   - 1–2 metres long.
2. If a farmer and his/her family, or urban household, can collect more compost making materials, the pit can be made longer, but NOT either wider or deeper.
3. If a pit is deeper than 1 metre, the material at the bottom does not get decomposed because many of the micro-organisms cannot live so deep down as the oxygen they need will not reach them.
4. Before any materials are put into the pit, the sides and bottom should be checked to make sure no water is leaking into the pit.
5. The bottom and sides should be plastered with a mixture of fresh animal dung and water, or top soil and water, to seal the surface so that the moisture in the compost materials is kept in the pit.

Filling the pit

The foundation layer

1. Dry plant materials, e.g. strong straw, stalks of maize and sorghum or tall grasses, as well as rose and hedge clippings from gardens, are used for the foundation. These need to be crushed or chopped or broken into short lengths (about 10–15 cm). If possible, let the domestic animals walk over them and sleep on them for one or two nights so the materials get broken up and mixed with urine and dung.
2. Spread the materials evenly over the bottom of the pit to make a layer 15–25 cm thick. Then sprinkle/scatter water evenly so that the materials are moist, but not wet.
3. This is a very important layer in making compost in a pit as it makes sure that air can circulate to the bottom.
4. Cover the foundation layer with large leaves, e.g. those of pumpkin, banana, fig leaves etc., and/or plastic to keep the material moist.

Putting the other layers into the pit

1. Each week, collect materials and put them in a container such as an old jerry can or pile them next to the compost pit. Mix the fresh moist materials with dry ones. These materials can come from spoiled animal feed, old animal bedding, from cleaning the house and compound, preparing vegetables for food, clearing paths, weeding, stems and leaves after harvesting vegetables, damaged fruits and vegetables, etc.
2. If the farmer has a biogas digester, the bioslurry can be collected also to be mixed with the other materials. The bioslurry is an excellent compost making aid.
3. At the end of a week, remove the large leaves or plastic covering the top of the foundation layer so they can be used again, or leave the leaves to become part of the compost if they are too damaged to be used again.

4. Make a mixture of compost making aids like good soil, manure and/or some ashes as a fine powder. Mix these with the dry plant material, or with the mixture of dry and moist plant material.

5. Add the prepared composting materials in layers. Each layer is 15–25 cm thick at the edge and a bit thicker in the middle so that the heap becomes dome-shaped (see Figure 6).

6. Cover each of the layers with a thin layer of soil and/or big leaves like those from banana or pumpkin or fig trees so that the composting materials do not dry out.

7. Repeat this process each week, or whenever there are enough materials collected to make one or two new layers, until the pit is full. Make the centre of the layers in the pit higher than the sides so that the top has a dome shape.

8. Put a ventilation and/or testing stick into the middle of the pit.

Making the covering layer

The pit full of composting materials needs to be protected from drying out, and also from animals disturbing it.

1. The covering layer should be made of mud plaster, with or without cow dung, with only the ventilation stick coming out of the top. It is then covered with wide leaves of pumpkin, banana, fig trees, etc. or plastic can also be used to protect the top of the pit. The leaves or plastic:
   - Prevent rainwater from getting inside the pit; and
   - Help keep heat inside the pit.

2. The compost pit can be left untouched until there is mature compost inside it, or it can be turned over and checked for the progress in decomposition. The top of the pit will sink down as the compost materials get decomposed. However, if the compost is turned over, it will lose moisture. So, it is best only to turn compost over if there is enough water and/or urine to make it moist again while it is being turned over.

3. The process for turning over the compost from the pit is the same as that described for the Indore pit method.

4. In a warm climate, mature compost can be made in three to four months. In colder climates, decomposition can take six months or a year.

5. The mature compost can be left covered and stored in the pit until it is needed for adding to the soil.

TRENCH COMPOSTING

Trench composting is suitable for groups. These can be groups of farming households, environmental clubs in schools, or youth group members who agree to work together to collect the materials, make the compost, and then share it among the members, or use it in their common garden.

Trench composting is good for mixed groups of men and women because men can do the heavy work of digging the trench and turning the compost materials, while the women can contribute materials and help carry the mature compost to where it is needed, including their own fields and gardens.

1. Plan to make compost in a trench at the end of the rainy season when there is plenty of suitable compost making materials available from clearing paths and compounds, etc., so that the mature compost is ready for the next growing season, or for making nursery beds for raising tree and vegetable seedlings.

2. The trench should be made at a convenient place for the members of the group to bring the collected materials; for example, near a path used by the members. It should also be under the shade of a tree to protect the people working to make the compost from getting too hot in the sun. In some communities, the people making and turning the compost do it in the evening or even at night to prevent getting overheated. The strong smell that can come from decomposing materials is also reduced in the cool of the evening or night.

3. A good size for the trench is as follows:
   - 0.5–1 metre deep, but not deeper than 1 metre;
   - 1–1.5 metres wide; and
   - 2.5 metres or longer if there are plenty of materials, even up to 10 metres long.

How to prepare and fill the trench

1. Mark out the size of the trench. *Note:* the length of the trench can be increased as more materials become available.

2. Dig down to 0.5–1 m and put the soil in a pile to one side of the trench. The soil is added in layers between the composting materials and/or used to cover the top of the filled trench.

3. The group members collect and bring materials from their houses, home compounds, cleaning paths, weeding, after harvesting vegetables, etc., if possible after having animals lie on the materials for one or more nights.

4. Look for and collect dry plant materials, such as long grasses and matting, sorghum and maize stalks to make a foundation layer. Get them broken up by animals walking and lying down on them. Put these materials as a bottom layer in the trench. Sprinkle/scatter water over the dry materials until they are moist, but not wet.

5. Mix all the collected materials together. Some or all of the following are suitable:
   - Cleanings from the house and from cooking;
   - Crop residues – leaves and stalks from harvesting and clearing/cleaning vegetable fields;
   - Grasses;
   - Chicken and goat and sheep droppings, cow dung; and
   - Ashes, etc.

   Add some old compost as a starter (like yeast).

6. Put the mixed materials in the trench in layers, each 20–25 cm thick at the sides and thicker in the middle.

7. Sprinkle/scatter water, or urine mixed with water over the materials, until they are moist but not wet. Any type of wastewater, even after washing clothes with hard washing soap, (but NOT with powder or liquid detergents) can be used for wetting.

8. Cover this layer with a thin layer of the soil taken from digging the trench.

9. Repeat this process of making layers until the trench is full and the middle is 25–50 cm higher than the surrounding ground.

10. Mix the soil that was dug out from the trench with straw, grasses, cow dung and water, in the same way as making a mud plaster to cover the walls of a house. Use this mixture to make a complete cover and seal over the top of the compost materials. Regularly check the mud plaster cover and repair cracks or other types of damage.

11. Put ventilation/testing sticks in the compost materials at about 1 metre intervals.

12. Finally, cover the trench with thatching grass or wide leaves of banana or pumpkin or fig trees, and/or plastic to keep in the moisture and heat.

13. Regularly use the testing sticks to monitor the progress of compost making.

14. The covered trench can be left untouched for three to four months, or longer, by which time mature compost will have been made. Evidence of compost making is seen first in the heat, and then in the fact that the heap shrinks down, and weeds start to grow on the mud cover.

How to turn over trench compost

1. After two months, the cover can be opened and the compost turned over. At the same time, the moisture balance and decomposition process can be checked. However, if the decomposition process is not complete, the compost will have a strong smell. It is best to do the turning over process during the early morning, or in the evening, or even at night to reduce the smell.

2. Turning over the compost is best done by digging out all the compost from about 50 cm at one end of the trench, and putting this outside the trench. Then the remaining compost is turned over in units of 50 cm into the trench so the materials at the top are put at the bottom and those at the bottom are put on top. The materials taken out from the first 50 cm strip are put back at the end of the trench.
This is the same method as that used in double digging a vegetable bed.

3. If the materials are not well decomposed and too dry, water can be sprinkled over the materials as they are turned over.

4. If the materials are too wet and smelling of ammonia, more dry materials can be added in the turning over process.
5. After turning over, the materials need to be covered and sealed as described above.

THE NADEP METHOD

The NADEP method is a development of the Indore method. It is named after its inventor, Narayanrao Pandaripade who was also called 'Nadepkaka'. This system is suitable for organized groups, such as growers associations, cooperatives, school environment clubs and youth groups, to make large quantities of high quality compost which they can use for themselves, or sell, for example, to vegetable growers, where high levels of nutrients are required.

The NADEP method produces nitrogen-rich compost using the least possible amount of cow dung. The system also minimizes problems from pests and diseases, and does not pollute the surrounding area because the compost is made in a closed tank.

After the NADEP tank has been filled with compost making materials and sealed, it is left for the decomposition process to take place without any further handling until the mature compost is required.

Selecting and preparing the site for the NADEP tank

The NADEP method uses a permanently built tank of mud or clay bricks, or cement blockettes. It is, therefore, important to choose the permanent site for the tank with care.

1. Select a site where there is enough space to collect the materials together before filling the tank, and where mature compost can be stored until it is needed.
2. The site needs to be near a source of water.
3. The site should be sheltered from rain, floods and wind. The best is in the shade of a tree, or on the north or west side of a building or wall. However, air must be able to circulate all round the tank.

FIGURE 7

A full NADEP tank built from bricks or blockettes

1m

2m

2–3 m

blockette or
mud brick

Building the NADEP tank

1. The inside dimensions of the tank are as follows:
   - Length: 3 metres
   - Width: 2 metres
   - Height: 1 metre
2. This size of tank requires 120–150 blockettes or mud bricks, four 50-kg bags of cement, and two boxes of sand. Five iron rods can be used to strengthen the floor, but they are not essential.

3. The building should be done by a properly qualified mason, i.e. someone who knows how to build such a structure.

4. The floor of the tank is made of bricks or blockettes laid on the ground and covered with a layer of cement.

5. Each of the four walls has three rows of holes or gaps between the bricks or blockettes, as shown in Figure 7.

6. After the tank is built, the walls and floor are covered with a light plaster of fresh cow dung mixed with water, and then the plaster is left to dry.

Filling the tank

A NADEP tank is filled in one or two days of hard work. It has to be done by a team. Before filling the tank, the following materials must be collected together:

1. Dry and green plant materials: 1 400–1 500 kg (or 14–15 sacks) are needed. Grass, hay or straw that has left over from feeding animals, or that has been damaged by rain, is very suitable.

2. Cow dung or partly dried bioslurry (the discharge from a biogas digester): 90–100 kg or 10 sacks.

3. Dried soil that has been collected from cattle pens, cleaning drains, paths, etc.: 1 750 kg are needed. The soil should be sieved to remove old tins, plastic, glass, stones, etc. Soil that contains cattle urine makes it very productive in the compost making process.

4. Water: the amount varies with the season and the proportion of dry to green plant materials available. However, usually an equivalent amount to plant materials is needed, i.e. 140–150 litres.

5. If urine from cattle and/or people is available, it should be diluted in the proportion of 1 part urine for 10 parts water (1 jug of urine put into 10 jugs of water in a bucket).

6. Before starting to fill the tank, the sides and floor of the tank are thoroughly wetted with slurry made from fresh cow dung mixed into water.

7. The three layers used to fill the tank are as follows:

- **First layer:** use 100–150 kg of dry or mixed dry and green plant materials to make a layer 15–25 cm thick at the sides, and slightly thicker in the middle.
- **Second layer:** Mix 4 kg of cow dung or 10 kg of fresh biogas slurry in 25–50 litres of water and sprinkle or scatter it over the plant materials so they get completely moistened.
- **Third layer:** Cover the wet plant waste and cow dung or slurry layer with 50–60 kg of clean, sieved top soil.

8. Continue to fill the tank like a sandwich with these three layers put in sequence. Put more materials in the middle of the tank than around the sides. This will give a dome shape to the filled tank with the centre 30–50 cm higher than the sides (see Figure 7).

9. Cover the last layer of plant materials with a layer of soil 7–8 cm thick. Make a cow dung plaster and cover the soil so that there are no cracks showing. The top of the filled tank can also be covered with plastic, particularly to protect the compost making process during rainy seasons.

10. After the tank is filled, the progress of compost making can be tested by pushing a stick into the tank through the gaps in the wall. In a school or agricultural college, the students can monitor the changes in temperature by inserting a long thermometer, e.g. a soil thermometer.

11. As the materials decompose in the compost making process, the top of the filled tank will shrink down below the sides of the tank.

Following up on the NADEP compost making process

It is important to keep the contents of the tank moist, i.e. with a moisture content of 15–20 percent.

1. Check the mud plaster seal on the top of the tank and fill any cracks that appear with cow dung plaster.

2. Pull out any weeds if they start to grow on the surface, as their root systems can damage the cover and take water out of the compost.

3. If the atmosphere gets very dry and hot, such as in the dry season, water can be sprayed through the gaps in the walls of the tank.

The decomposition process for compost to be made takes about three to four months in a warm climate. When it is mature, it is dark brown, moist, and with a pleasant earthy smell: little can be seen of the original materials that were put into the tank.

This mature compost should not be allowed to dry out or it will lose a lot of its nitrogen. However, before the compost is mixed to make nursery soil, it should be sieved. The sieved compost is used in making the soil for the nursery beds, and the remainder is kept and added to a new compost-making process. One NADEP tank of the size described here can produce about 300 tonnes of high quality compost.

FOLLOWING UP ON CONDITIONS IN THE COMPOST MAKING PROCESS

When the compost pit has been filled or the piling of materials is complete, it should be checked regularly to make sure that there is enough but not too much moisture, and that it is getting hot, at least in the first two to three weeks.

For compost made by piling materials on the ground:

- The stick can be inserted or pushed in horizontally between two layers about half way up the pit; or
- The stick can be pushed in vertically in the centre of the heap so it goes through all the layers. However, it is best if the stick or length of bamboo is placed in the centre after the foundation layer has been laid and then the layering process is completed with the stick remaining vertical.
- The stick must be longer than the height of the heap so that it can be pulled out and examined.

For compost made in a pit:

- The stick or length of bamboo is pushed in vertically through the whole layer, or put in place while the compost pit is being filled.
- The stick must be longer than the depth of the pit.

Checking heat and moisture

One week after all the materials have been put in a heap or pit, and it has been covered, remove the inserted stick and immediately place it on the back of your hand.

1. If the stick feels warm or hot and the smell is good, the temperature is normal for the compost and good decomposition has started.

2. If the stick feels cool or cold and there is little smell, the temperature is too low for good composition. This usually means that the materials are too dry, and some water and/or urine should be added.

3. If the stick is warm and wet, and there is a bad smell like ammonia, this indicates that there is too little air and too much water in the compost. The materials will be rotting and not making good compost.

Correcting the problems
If the materials are cool and dry

1. Lift up the top layers and put them to the side of the pit or heap.

2. Sprinkle water or cattle urine or cattle urine diluted with water on the material in the bottom.

3. Then put back the material in layers of about 25 cm each sprinkling water or a mixture of water and urine over each.

4. Replace the testing stick and cover the heap or top of the pit with soil, leaves, plastic etc., as described earlier.

If the materials are too wet

1. Collect some more dry plant materials and/or some old dry compost. Break up and mix the materials. If old dry compost is not available, use only the dry plant materials.

2. Lift off the top of the heap or take out the top half of the materials from the pit and put them to one side.

3. Mix the new dry materials with the wet compost materials in the bottom.

4. Put back the materials from the side of the heap or pit. If these materials are wet and decaying, put in alternate layers of new dry plant materials with the wet materials.

5. If the top materials are moist and brown showing compost making has started, put them back as they are.

6. Put back the vertical testing stick.

7. Do NOT seal the top but make a new test after a week.

   o If the stick is warm or hot and the smell is good, good compost making has started and the heap or top of the pit can be sealed and covered.

   o Testing for heat and moisture should be done every week to 10 days until mature compost is made.

QUALITIES AND USE OF GOOD COMPOST

Although the quality of compost is best evaluated through the growth and productivity of the plants grown on soil treated with it, it is possible to evaluate compost quality through seeing, touching and smelling:

o Good quality compost is rich in plant nutrients and has a crumb-like structure, like broken up bread.

o It is black or dark brown and easily holds moisture, i.e. water stays in it, and it does not dry out fast.

o It has a good smell, like clean newly-ploughed soil, with a smell somewhat like that of lime or lemon.

Using compost

- Mature compost is best stored in a pit or heap until it is needed. If it is kept dry and covered, mature compost can be stored for several weeks without deteriorating.

- The stored mature compost should be kept in a sheltered place, e.g. under the shade of a tree or in a shed, and covered with leaves and/or soil and sticks to prevent the nutrients escaping to the atmosphere, and animals trampling on and damaging the mature compost heap.

- Mature compost should be taken to the field early in the morning or late in the afternoon.

- For crops sown by broadcasting, the compost should be spread equally over the field, or the part of the field chosen to be treated with compost. The compost should be ploughed in immediately to mix it with the soil and prevent loss of nutrients from exposure to the sun and wind.

- For row planted crops, e.g. maize, sorghum and vegetables, the compost can be put along the row with the seeds or seedlings.

- For trees, compost is put in the bottom of the planting hole and covered by some soil when the seedling is planted out. It can also be dug into the soil around the bottom of a tree seedling after it has been planted.

- Time and effort are needed to make good compost, so it is worthwhile to also put in time and effort into using it properly in the field.

Problems in using compost

Improper use: The aim of preparing compost is to increase soil fertility and crop yields. Sometimes, a farmer will try and spread a small amount of compost over a wide area, and then be disappointed when he/she does not see any improvements to his/her soil and crops. If only a small amount of compost has been made, it is best to put it on a small area of land than spread it thinly over a wide area. Every farmer must aim to produce enough compost for her/his particular farmland to get a better yield (return). A guide on how much to add is given in the next section. Compost should not be left exposed to sun and wind on the surface of the soil, but buried immediately. Compost should not be added to empty fields. This is a waste of time and effort. By the time the crop gets sown, the compost will have lost a lot of its nutrients and make the farmer disappointed with his/her effort to make and use compost.

Carrying compost: Compost is bulky. For best results a farmers needs to carry up to 30 to 70 sacks (3–7 tonnes) of compost to cover a one-hectare field.

BOX 3
FARMERS SOLVING THE PROBLEM OF CARRYING COMPOST

The farmers of Adi Abo Mossa near Lake Hashengi in Ofla of Tigray and in Gimbichu of Oromiya Regions have solved this problem by using their donkeys to carry the sacks containing mature compost from the compost pit to the field. Other farmers, in Adi Nefas, have organized the making of compost to be near their fields so they only need to carry the mature compost a short distance. If farmers are seriously convinced about the usefulness of compost, they find their own ways to solve these problems.

AMOUNT OF COMPOST FOR ONE HECTARE

Where soil fertility has been lost through many years of land degradation, 100–150 kg of chemical fertilizer, such as urea and DAP recommended by the Ethiopian government, can improve the yield, often dramatically if there is enough rain, in just one year. However, the effects of chemical fertilizer last for only one growing season, so it has to be added every year.

One tonne of compost is not enough to get the same increase in yield immediately in the year it is added to the soil for the first time. This is because the amount of main plant nutrients (NPK) found in one tonne of compost is lower than those in 100 kg of chemical fertilizer. However, the effects of compost last for two or more growing seasons.

In Europe, where the soils are cold and there is much rain, the general guideline is that 20–25 tonnes of compost are needed to replace 100–150 kg of chemical fertilizer. The range is because the nutrient content of compost depends on the materials used to make the compost. Compost made only with plant materials usually has a lower nutrient content than compost made by including animal dung and urine.

In Ethiopia, more research is needed to find out how much compost is needed to get good yields in the different agro-ecological zones of the country. However, in Tigray, it has been found that compost added at the rate of 3.2–6 tonnes per hectare can give greatly improved yields, which are as good as, if not better than those from chemical fertilizer (see Edwards *et al.*, 2007). In wetter areas, farmers find that 5–8 tonnes per hectare can improve crop yields.

A rough guide on amounts of compost that can be made and used by farmers in Ethiopia

The following is a guide on the amount of compost to aim to produce under different environmental conditions:

- Mature compost to give a rate of 8–10 tonnes per hectare can be achieved in areas where there are plenty of composting materials, a good water supply and labour. Farmers working in groups are more likely to be able to produce large quantities of good quality compost than farmers working alone. These quantities have been achieved in Adi Abo Mossa village in Southern Tigray and Gimbichu district in Oromiya Regions.
- Mature compost to give a rate of around 6 tonnes per hectare can be achieved where there are medium amounts of composting materials, and water and labour

are available. These quantities have been achieved by farmers working in Central Tigray near the town of Axum.

- Mature compost to give a rate of around 3.5 tonnes per hectare can bring improved yields. This can be achieved even in areas with low availability of composting materials, as long as there is enough water to moisten the composting materials. These rates have been achieved by farmers in the semi-arid eastern parts of Tigray.

- Where there are only small amounts of composting materials, e.g. for farmers who have very small plots of land and for women-headed households, working together to fill a common pit can make better quality compost than working alone.

- If only a small quantity of compost is made, it is important to apply it properly to a small piece of land to make it as useful as possible, instead of spreading it thinly over a wider area.

- Soil given compost in one year will not need it again in the next year as the good effects last for more than one growing season. The new compost can then be used for the part of the field that had no compost the previous year. Farmers that are able to apply the equivalent of 8–10 tonnes per hectare say that the good effects last for up to three years.

Planning to make good quality compost

The following factors need to be discussed and decisions made in order to prepare enough compost for a chosen piece of land:

- Making and using compost correctly needs labour, so farmers and their families have to be prepared to work hard to make good compost, and get good results from using it correctly.

- Every farmer, agricultural agent, supervisor and expert working in the area should be convinced about the use and importance of compost. If everyone is convinced, then all will be willing to work hard to get good results.

- Farmers and their families need to identify the materials in their fields, compounds and surroundings that can be used to make compost.

- The pit or pits for making compost should be near the source of materials, like the edge of a field for weeds and crop residues, or inside or near the family compound for waste from the house, home garden and animal pens.

- Farmers living near small towns and villages may be able to arrange to collect waste materials from the houses, hotels, and other institutions where food is made in the town or village, or have the waste materials brought to an area convenient for a group to make compost together.

- The youth in a village or small town can be trained to make compost from the wastes in the town, and to use it to grow their own vegetables or sell it to the farmers.

- Farmers should work with their development agents, supervisors, experts, and other persons to help them make decisions about how to make compost depending on the local availability of composting materials, the place where the compost is to be made and the fields where it is going to be used to improve the soil and crop yields.

REFERENCES AND FURTHER READING

Arefaine Asmelash. 1994 EC/2002 GC. *Dekuaie tefetro: intayn bkhmeyn* (Making compost: what it is, and how is it made). In Tigrinya. Tigray Bureau of Agriculture and Natural Resources and Institute for Sustainable Development, Addis Ababa, Ethiopia.

Assefa Mitiku Tegegn & Sisay Kebede Haile. 1991 EC/1999 GC. *Yebsebashe Azegedgadget leanesetenya geberyewotch* (Compost making for smallholder farmers). In Amharic. AgriService, Addis Ababa, Ethiopia.

Brandjes, P., P. van Dongen & A. van der Veer, 1995. *Green manuring and other forms of soil improvement in the tropics.* (English translation by H Huf.) Agrodok-series No. 28. Agromisa, Wageningen, The Netherlands. Distributed by CTA (Technical Centre for Agricultural and Rural Cooperation, ACP/EEC).

Dalzell, N,A. & S. Riddlestone. 1987. *Soil management: Compost production and use in tropical and subtropical environments.* Bernan Associates, USA.

Edwards, S., Arefayne Asmelash, Hailu Araya & Tewolde Berhan Gebre Egziabher, 2007. *Impact of compost use on crop yields in Tigray, Ethiopia, 2000-2006 inclusive.* FAO, Rome. Available at: http://www.fao.org/documents/pub_dett.asp?lang=en&pub_id=237605

Gershuny, G. & D.L. Martin (eds). 1992. *The Rodale book of composting: Easy methods for every gardener.* Rodale Press & St Martin's Press, USA.

Howard, A. 1940. *An agricultural testament.* Seventh Impression 1956. The Other India Press, Goa, in association with Earthcare Books, Bombay and Third World Network, Penang.

Inckel, M., P. de Smet, T. Tersmette & T. Veldamp, 1980. Revised by M. Louis & M. Leijdens, 1999. *The preparation and use of compost.* Agrodok-series No. 8. Agromisa, Wageningen, The Netherlands. Distributed by CTA (Technical Centre for Agricultural and Rural cooperation, ACP/EEC).

Roulac, J. 1996. *Backyard composting.* Green Earth Books, Totnes, England.

Vasanthrao Bombatkar (1996). *The miracle called compost* (An introduction to the art of making NADEP compost). (Translated into English by Smt Sharada Swamy & Dr S M Ramachandra Swamy, edited by Norman Dantas.) The Other India Press, Mapusa Goa, India.

Design | Layout | Illustrations
Pietro Bartoleschi and Arianna Guida
studio@bartoleschi.com
Printed in Italy on ecological paper, 2011